QUALITY ASSURANCE OF AQUATIC FOODS

The Textbook on "Quality Assurance of Aquatic Foods" deals with the quality and safety of aquatic food. It focuses clearly on biological and chemical hazards, antibiotic and pesticide residues, and heavy metal contaminants associated with aquatic food. The quality problems in various aquatic food products and their methods of assessment are exhaustively dealt in this book. Besides, food quality management systems like HACCP, SSOP, SCP, GHP, GMP have also been explained for easy adoption. The International and National Standards prescribed by FSSAI, EIC, BIS, Codex, USFDA, ISO and EU for aquatic food products are explicitly given. This book is also useful to the personnel of aquatic food industries to improve their working knowledge on maintaining the quality and safety aquatic foods.

Prof. G. Jeyasekaran, Director of Research of Tamil Nadu Dr. J. Jayalalithaa Fisheries University, has been working for more than 33 years and involved in teaching, and research activities pertaining to fish quality, safety, authenticity, microbiology and biotechnology. Being the first Director of Research of the University, he was involved in the overall development of the research programmes.

Prof. R. Jeya Shakila, Head of the Department of Fish Quality Assurance and Management, Fisheries College & Research Institute, Tamil Nadu Dr. J. Jayalalithaa Fisheries University has been working in the fields of fish biochemistry, quality, safety and authenticity for more than 21 years. She worked in Defence Food Research Laboratory (DFRL), Mysore for her Ph.D. degree in the area of Biogenic amines from fish and fishery products.

QUALITY ASSURANCE OF AQUATIC FOODS

Prof. G. JEYASEKARAN
and
Prof. R. JEYA SHAKILA
Department of Fish Quality Assurance and Management
Tamil Nadu Dr. J. Jayalalithaa Fisheries University
Thoothukudi 628 008, Tamil Nadu, INDIA

NARENDRA PUBLISHING HOUSE
DELHI (INDIA)

First published 2021
by CRC Press
2 Park Square, Milton Park, Abingdon, Oxon, OX14 4RN
and by CRC Press
6000 Broken Sound Parkway NW, Suite 300, Boca Raton, FL 33487-2742

© 2021 Narendra Publishing House

CRC Press is an imprint of Informa UK Limited

The right of G. Jeyasekaran and R. Jeya Shakila to be identified as authors of this work has been asserted by them in accordance with sections 77 and 78 of the Copyright, Designs and Patents Act 1988.

Print edition not for sale in South Asia (India, Sri Lanka, Nepal, Bangladesh, Pakistan or Bhutan).

British Library Cataloguing-in-Publication Data
A catalogue record for this book is available from the British Library

Library of Congress Cataloging-in-Publication Data
A catalog record has been requested

ISBN: 978-0-367-61939-8 (hbk)
ISBN: 978-1-003-10717-0 (ebk)

**NARENDRA PUBLISHING
HOUSE
DELHI (INDIA)**

CONTENTS

PREFACE

Since aquatic food is considered as the best muscle food having health beneficial nutrients like omega-3 fattyacids, essential aminoacids, vitamins and minerals, the protein malnutrition among people in the World, particularly from the regions of developing countries, can be overcome by the supply and consumption of quality and safe fish and fishery products. Fish Quality Assurance and Management is an important area under the fish processing sector. The passage of Food Safety and Standards Act and the formation of Food Safety and Standards Authority of India (FSSAI) by the Indian Parliament underline the importance of producing safe food for human consumption. The Govt. of India also recently passed Food Security Act, which emphasize the necessity for providing nutritional food to all the people in India. All these developments lead to the enormous need for having highly technically qualified manpower in the field of Fish Quality Assurance and Management. Intensive research is needed and to be carried out in the SAUs and Research Institutes, for the production of quality and safe fish and fishery products so as to meet the requirements of national and international quality standards being prescribed by regulatory agencies of different Governments at national and international levels. This specialization of Fish Quality Assurance and Management is evolved from the field of Fish Processing Technology as that of Aquatic Animal Health and Management from the field of Aquaculture.

It is a fact that Food Regulatory Authorities need specialized technical manpower to verify food safety management systems (FSMS) in fish processing establishments, and technically qualified personnel to ensure supply of quality fish products to the consumers by producers and retailers, Seafood processing industries need qualified personnel to implement FSMS in their establishments, and Central and State Governments need manpower with specialized knowledge in fish quality and safety to combat with new and emerging safety issues in aquaculture sector, and Fish quality inspectors need quick and rapid kits for on-site monitoring fish quality and safety in fish landing centers, retail stores and supermarkets to check fraudulence practices.

India fetches a considerable amount of foreign exchange through the export of aquatic food. The fish processing industries are recently facing severe competition at the international markets due to stringent quality control measures adopted by different seafood importing countries. This makes India to implement total quality assurance and management systems such as hazard analysis critical control point (HACCP) system, European Union (EU) hygienic regulations and ISO 9000 standards on aquatic food for meeting the ever-increasing quality demands at the international markets. However, new kinds of quality problems such as antibiotic residues in farmed shrimps also emerge in the seafood sector. In order to control such quality problems, the Scientists of fish post-harvest technology particularly working on aquatic food quality and safety are finding ways and means through research. With a view to disseminate the knowledge acquired by us on various aspects of aquatic food quality and safety to the students and staff of fisheries institutions, research organizations, quality control personnel of fish processing industries for effectively controlling the quality and safety of aquatic food, this textbook is published.

We strongly feel that this book will be of immense useful to the personnel involved in the field of aquatic food processing technology particularly in aquatic food quality assurance.

Authors

CHAPTER 1

━━━━●━━━━

INTRODUCTION

Aquatic food has been a popular food globally in many countries and in few countries it constituted the main supply of animal protein. The low fat content of aquatic food and effects of w-3 polyunsaturated fatty acids found in fatty fishes on coronary heart diseases are extremely important aspects for human beings particularly in developed countries. However, the consumption of aquatic foods may also lead to diseases due to infection or intoxication. Some of the diseases have been specifically associated with consumption of aquatic food. The quality and safety of aquatic food are of major concerns to public health authorities and aquatic food processors. It has been estimated that there are about 600 million cases per annum of food-borne illnesses and that the cost of these illnesses is in the order of many billions of dollars per year.

Aquatic food quality assurance is a major issue presently facing the aquatic food industry around the World. Aquatic food safety is a prerequisite for protecting consumer health, which also serves the interests of producers and those involved in processing and marketing of aquatic food products. Aquatic food quality assurance includes all activities carried out to ensure the quality and safety of aquatic food. Every stage from initial handling to processing, storage, distribution and consumption must be included in a aquatic food quality assurance programmes. The overall aim is to provide a systematic approach to all control and inspection activities through a managed programme based on proper scientific principles and appropriate risk assessment, leading to careful targeting of inspection and control resources. Aquatic foods are in the forefront of food safety and quality improvement programmes as they are among the most internationally traded food commodities. About 50% of aquatic food in the International trade originated from the developing countries such as India.

The true incidence of diseases transmitted by aquatic food is not known. There are many reasons for this kind of situation. In most of the countries, there is no obligation to report on food-borne diseases to public health authorities. In some countries, which have a reporting system, there is severe under-reporting. It has been estimated that as few as 1% of the actual cases of food-borne diseases are recorded. This is because neither the victim nor the clinician is aware of the etiological role of foods. Furthermore, the food responsible is often not available for analysis and the true vehicle for the disease agent is not identified. Of the food-borne disease outbreaks reported, aquatic food was the food most frequently associated with the disease. Diseases associated with various types of aquatic food so far reported indicated that finfish was most frequently involved followed by bivalve molluscans, and crustaceans. A considerable number of disease outbreaks related to finfish recorded were of unknown etiology. Most common were intoxications related to biotoxins and histamine, which accounted for about 65% of all recorded outbreaks. Various bacteria, virus, parasites and chemicals caused the rest of the diseases associated with the consumption of aquatic foods. The outbreaks of aquatic food borne diseases related to consumption of molluscans were mainly due to the contamination with viruses. However, the outbreaks of diseases due to the consumption of crustaceans were related to pathogenic bacteria.

India has a long coastal line of 8118 km and an economic zone of 2.02 million sq.km, which offer a huge resource for aquatic foods. The total Indian catch has been generally increasing every year. The export of aquatic food from India has brought a huge sum of foreign exchange to a value of about Rs. 45,106/- crore during 2017-18. Despite that, the contribution of India to global aquatic food export is not significant. This was mainly because of the poor image abroad regarding the quality of its product. The processors and exporters should ensure quality from the stage of raw material to final packaging. Similar safety / quality regulations have also to be ensured in the aquaculture industry from the stage of seeds to the final product. In order to further increase the export, there is an urgent need to consistently assure the quality and safety of aquatic food at the International level.

CHAPTER 2

———•———

QUALITY OF AQUATIC FOOD

Quality has been commonly thought of as degree of excellence. In general terms, quality is defined as the composite of those characteristics that differentiate individual units of a product and have significance in determining the acceptability of that unit by the buyer. In regard to aquatic food products, it is a concept based on the composite of several characteristics such as physical, chemical, biochemical, organoleptic and bacteriological. Thus, quality embraces intrinsic composition, nutritive value, degree of spoilage, damage, deterioration during processing, storage, distribution and sale to the consumer, hazard to health, satisfaction on buying and consuming, aesthetic considerations, yield and profitability to the producer and seller. In short, aquatic food quality means all that attributes which consciously or unconsciously the aquatic food consumer or buyer considers should be present. The decisions about what constitutes quality rest ultimately with the aquatic food consumers. The producer, who wishes to know what aspects of quality are important be in close liaison with the customers, reacts quickly to their changes in taste and keeps systematic records of their reactions to or complaints about the products. It is equally necessary to be aware of the economic factors that affect attitudes to quality such as cost, supply and demand. Some improvements in quality or the achievement of consistent quality can be effected at no extra cost, but some cannot be done without increasing the cost. Extra costs must be balanced against the long-term benefits of continued customers' loyalty and of increased sales.

Quality is broadly grouped as intrinsic and extrinsic, based on the origin of the quality factors. Intrinsic quality is the sum of attributes that are inherent in the raw material. Extrinsic quality is the sum of the effects of all the treatments fish receive after catch till they reach the consumer. It is obvious that intrinsic quality is beyond direct control, while extrinsic quality can be properly controlled. Quality control (QC) is defined as the operational techniques and activities that are used to fulfill quality requirements. Quality assurance (QA) is defined as all activities and functions concerned with the attainment of quality. Quality management (QM) is often used interchangeably with quality assurance. In the food industry, the term quality management has been used to focus mostly on the management of the technical aspects of quality in a company. Total quality management (TQM) is an organization's management approach, centered on quality and based on the participation of all its members and aimed at long-term success through customer satisfaction and benefits to the members of the organization and to society. The term "Food Safety Objective" (FSO) translates food safety risk into a measurable goal and is expressed as the concentration or frequency of a hazard in a food, at the point of consumption, that is considered safe or meeting the level of protection set by society.

2.1. INTRINSIC QUALITY

It is the inherent quality. Several factors affect the intrinsic quality. They are species, size, sex, condition and composition, parasites and other organisms, naturally toxic fish, contamination with pollutants and occasional peculiarities.

1. Species

Certain fishes are generally costly because they are rated as 'good quality' fishes. In India, seerfish and pomfret are costly and hence, rated as good quality fish. In this case, the quality is related to species. Hence, species identification assumes importance where the external features of the fish are destroyed while processing as in the case of frozen fillets or minced meat.

2. Size

Generally large specimens of a given species fetch better price. Large sized shrimps, crabs and lobsters fetch higher price in the markets. Processors place higher value on large specimens because of higher percentage yield of edible material, lower handling costs per unit weight, better keeping qualities and yielding of more uniform products. But for certain purposes, the biggest size is not the optimum as in the case of canned shrimps because of difficulties in controlling fill weights. Control of size is to some extent exercised by choosing the fishing grounds, seasons or methods. Manual or mechanical sorting is also possible after catch.

3. Sex

Preference of fish based on sex is known to exist, but there is no documentary evidence of this, where Indian waters fish species are concerned. The females of certain species such as Sturgeon are more valued in Europe probably for their roe or caviar.

4. Condition and composition

Condition refers to the biochemical and physiological state. In certain seasons, fish is said to be in good condition and at other times in poor condition. Fish in poor condition is known to have more water content and less fat content. This is usually so after spawning. Sardines and mackerels are known to show wide variation in their fat content depending on season. Sardines in October, November and December have very high fat content and are preferred by many during this season. Feedy fish are known to be prone to belly bursting and hence they cannot be kept for long. Glycogen content in certain shellfishes varies with season. When glycogen is lower, the amount of lactic acid formed is also lower and hence the pH higher. This means that such fish will subject to microbial attack earlier. Crabs are known to differ in their meat content dependent on full moon or new moon periods. During new moon period, their meat content is higher. A special problem in certain fishes is 'chalkiness' i.e. the fish flesh becomes more white during storage and texture becomes poor. This occurs in seerfish during certain seasons and is related to more acidic pH attained. Considering many of these points, regulations exist in some nations on fishing seasons. Low post - mortem pH also leads to 'gaping'. i.e. the tendency of fish fillets to split into fissures and holes.

5. Parasites and other organisms

A parasite is an organism living on or inside another and depending upon it for some of its vital needs, particularly nutrient. Incidence of parasites in fish is another quality factor. These parasites are not easily visible and can cause difficulties when discovered by the consumer. Only few of these parasites are harmful. Many of these parasites are located in the head, viscera, etc. which are not consumed. Though thorough cooking kills all parasites and renders them completely harmless, these parasites cause health problems where fish is consumed raw. Some parasites are seen in tuna as also in frog legs. The parasites in fish mostly belong protozoa, flat-worms (platyhelminths), round worms (nematodes) and certain crustacea. Bacteria and fungi also cause disfiguring diseases in fish. Protozoan such as myxosporidian infection may cause softening of fish eg. *Chloromyscium thyrsites*. Many flat worms are free living but flukes (trematodes) and tape worms (cestodes) are parasites. Diseases in human beings are caused by consuming the cysts of two types of flat worms, the lung fluke (*Paragonimus* sp.) and the broad tapeworm (*Diphyllobothrium latum*). Round worms (nematodes) may be found in the gut and viscera of fish or encysted in the flesh. The cod worm (*Porrocaecum decipiens*) is a well-known example and other example is *Anisakis* sp., which is fairly common in squids and herrings. The occurrence of crustacea, fungi and bacteria is not a major problem in processing. Crustacea contain many parasitic forms, mostly belonging to the sub-group known as copepods. The sea louse, *Lepeophtheirus* sp. that is found on fresh run salmon and sea trout, is an obvious crustacean. Fungal and bacterial infections are widespread and can be of particular importance in fish farming and so only few are of importance for the fish processors. An example of a fungal disease is the condition known in the UK as 'greasy haddock'. It is caused by *Ichthyosporidium hoferi*. The flesh of infected specimens is soft and has a somewhat sweetish, slightly sickly smell. When such fish is smoked, white spots are seen in the flesh. Lesions, nodules (small lump or swelling) and pustula (pimple or blister) areas found in the commercially caught fish may be attributable to bacteria, mainly *Aeromonas* sp.

6. Naturally toxic fish

Some fishes are naturally toxic and cause injury to health or even death. Fish and shellfish poisonings are due to toxins in the flesh of animals either intrinsically present or derived from the food they consume. This is true for puffer fish poisoning (caused by tetradotoxin), ciguatera (caused by cigua toxin and possibly other toxins) and paralytic shellfish poisoning (caused by saxi toxin and other toxins). Quality control personnel should be able to identify such species. Ciguatera poisoning occurs by consuming certain carnivorous fish such as barracudas, snappers, etc. harvested from shallow waters in or near tropical and subtropical coral reefs at certain seasons. The ciguatoxin (CTX) is responsible for the poisoning and the principal source is the benthic dinoflagellate, *Gambierdiscus toxicus*. It is rarely fatal. Cooking does not destroy the poison. Puffer fish poisoning is by consuming the puffer fish flesh contaminated with viscera. Mortality rate is 50% and largest incidence is reported from Japan. Tetradotoxin (TTX) is caused by the bacteria, *Vibrio alginolyticus*. Paralytic shellfish poisoning is caused by eating certain molluscs particularly mussels and clams. The toxicity is related to the occurrence of dinoflagellates such as *Gymnodium* sp. or *Ptychodiscus* sp. in the harvesting water. When these organisms are too much, the water assumes a reddish tinge and it is called as red tide phenomenon. The only protective measure is not to collect shellfishes from such waters. Paralytic shellfish poisoning (PSP) is also called either as mussel poison or mytilo toxin / clam poison or saxi toxin. Common sources of PSP are *Gonyaulax catenella, Alexandrium* sp. and *Pyrodinium* sp. The toxin responsible for diarrhetic shellfish poisoning (DSP) is okadic acid (OA), dinophysis toxin and the common source of DSP is *Dinophysis* and *Prorocentrum*. The common source of neurotoxic shellfish poisoning (NSP) is *Gymnodium breve* or *Ptychodiscus breve* and the toxin responsible is brevetoxins. Venerupin shellfish poisoning (VSP) is also called as oyster or asari poisoning and its common source is *Prorocentrum* sp. Amnesic shellfish poisoning (ASP) affects the brain leading to the loss of memory and the toxin responsible is domoic acid. ASP is the only shellfish poison produced by a diatom. Erythematous shellfish poisoning (ESP) affects the blood circulatory system by specifically attacking the red blood cells.

7. Contamination with pollutants

Contamination of fish with pollutants is another problem, which an ordinary fish consumer is unable to identify. The main classes of pollutants, which need to be considered, are metals and other elements, chlorinated hydrocarbons, mineral oils, radioactive isotopes and microorganisms. A large number of potentially harmful heavy metals and elements that include mercury, cadmium, lead, selenium and arsenic are known pollutants. The important persistent organic chemicals are a group of chlorinated hydrocarbons including the insecticides/pesticides DDT, aldrin, dieldrin, benzene hexachloride (BHC or lindane) and polychlorinated biphenyls (PCB's). The US Food and Drug Administration have imposed maximum permitted concentrations for DDT and dieldrin in most fish as 5.0 and 0.30 mg per kg wet weight, respectively. The residues of antibiotics such as chloramphenicol, nitrofurans, tetracyclines, etc. in the flesh of fish and shellfish are also dangerous to human beings. Very occasionally catches become grossly contaminated with chemicals, especially mineral oils resulting from accidental (during offshore oil exploration and oil tanker movements) or other kinds of large-scale release. Evidence of such contamination is apparent in tainted odour or flavour of the fish. Exposure to high energy radiation originating from radioactive isotopes can be extremely injurious to health. Naturally occurring radioactive isotopes occur in fish, as in other foods, but at very low concentrations not considered to have a significant effect on health. In some areas, artificial isotopes have been released in very small amounts into the environment.

Although most of the microorganisms present on the outer surface, gills and in the viscera of fish caught in unpolluted water are harmless to humans, food poisoning can be caused by two microorganisms that may occasionally occur. Pollution of fish by raw or inadequately treated sewage can, however, introduce pathogenic organisms, i.e. capable of causing health injury to humans. Sewage contains two types of faecal microorganisms such as bacteria and single member of the virus family. The bacteria include the large group of *Salmonella* and different members of which cause food poisoning, typhoid and paratyphoid and *Shigella*, which cause dysentery. The only virus known to be incriminated is that responsible for the severely disabling disease infectious hepatitis. These microorganisms are transferred via seawater and detritus to fish including crustaceans and molluscans living within several kilometers of sewage outfalls or sewage polluted rivers. No sewage microorganisms are normally detectable in the open sea. The molluscans harvested from the sewage-polluted areas cause health hazard if they are consumed raw, as they concentrate microorganisms because they are filter feeders. A large number of different types of faecal microorganisms and of pathogens are potentially present in any given sample but is far too expensive and time consuming to determine more than one or two of them. It is therefore a standard practice to select what is known as an indicator microorganism which is invariably present in sewage and whose behaviour in the environment and in the molluscans is to all intents and purposes indistinguishable from that of the pathogens. The indicator organisms now being generally determined are either a group known as faecal coliforms or the overwhelmingly predominant number of the group, *Escherichia coli*. Determinations of either indicator give very similar results.

8. Occasional peculiarities

Peculiarities like tumors, ulcers, nodules, abnormal colouration, abnormal odours, etc. are also rarely observed in fish. Physical damage to food fish caused by predators such as seal on salmon, shark on many species before capture is occasionally found. But, they do not pose a serious problem in processing. The flesh of cod and other gadoids is on certain occasions found to be in various shades of pink colour. This is due to the presence of the natural red carotenoid pigments such as astaxanthin and zeaxanthin believed to be derived from unusual feed or some metabolic disturbance. On oxidation, these red pigments turn into yellow colour. The most well-known of the peculiar odours and flavours is an odour and flavour reported in cod and other gadoids, mackerel and chum salmon. It is variously described as 'black berry, weedy, petrol, diesel, iodine and sulphide'. The substance responsible for this has been identified as dimethylsulphide (DMS). This does not occur normally in most fish, but may occur as result of feeding on certain organisms. Species of planktonic bivalve molluscs known as pteropods have been implicated, particularly *Spiratella helicina* or *Limacina helicina*. These pteropods contain dimethyl - β - propiothetin which is converted in the fish to DMS.

Freshwater fish including tilapia occasionally suffer from an earthy or muddy odour and flavour, which can reduce consumer acceptance. Earthy odour occurs particularly in fish caught in waters harbouring high concentrations of certain algae particularly blue green algae or microorganisms belonging to Actinomycetes that they have a similar odour. The earthy odour is probably caused by the presence of geosmin and related substances originating in algae or microorganisms present in the water. An iodoform like odour in shrimps is fairly common and can be objectionable in high concentration. High concentrations of geosmin (trans, 1-10-dimethyl – 1-9 decalol), a musty odorous compound, are found in the tail muscle of the shrimp grown in coastal culture ponds. It is responsible for the earthy-musty flavor in pond cultured penaeid shrimps.

2.2. EXTRINSIC QUALITY

It is due to the effects of treatment the fish receive after capture till it reaches the consumer. The important extrinsic quality factors are degree of freshness, conformity to the declared mode of presentation, weight, size, ingredients and food additives, acceptability of the processing methods and suitability of containers and packaging methods. Special significance is attached to the cleanliness and sanitary conditions under which the products are handled and processed as evidenced by the degree of contamination with pathogenic and faecal indicator organisms.

Quality deterioration and extrinsic quality defects in raw material

Quality deterioration means that those natural processes of quality reduction, which occur after harvesting and are quite independent of deliberate intervention of humans, whereas extrinsic defects are quality reductions in the post-harvested material caused by deliberate or accidental actions of humans. There is a possibility of a considerable degree of planned control over both the quality deterioration and quality defects. In the aquatic food industry, it is sometimes impossible to draw a clear distinction between raw material and products, because, what is raw material to a processor may be finished product to a retailer. The raw material is considered as chilled or unchilled whole fish and shellfish, gutted or ungutted fish, and unpeeled or unshucked raw shellfish, and in every case it refers to raw uncooked material. The condition of such raw material is basic to the quality of many products. The causes of quality deterioration are broadly classified into microbiological and non-microbiological.

1. Microbiological deteriorations

Microorganisms are present on the external surfaces including slime and in the gills and gut of fish. But, they are kept away from invading the sterile flesh by the normal defence mechanisms of fish, when they are alive. On death, the microorganisms, or the enzymes they secrete are free to invade or diffuse into the flesh where they react with the complex mixture of natural substances present resulting in a well-defined sequence of changes in odoriferous and flavourous compounds. Initially compounds having sour, grassy, fruity or acidic odours are formed. Later, bitterness and sulphide odours appear. Finally, in the putrid state, the odour is ammoniacal and faecal. Many marine species contain the odourless compound, TMAO at a concentration of 0.2% to 2%, and this by the reaction of bacteria is reduced to TMA, which possibly in conjunction with fatty substances is alleged to smell 'fishy'. But, certainly on its own, it is always recognized as being ammoniacal. Elasmobranchs contain high concentrations of urea, which is microbiologically degraded to ammonia. At later stages of spoilage, microorganisms through their secreted proteolytic enzymes also attack the structural components, proteins resulting in a gradual softening of the flesh. Occasionally a different sequence of changes occurs where storage conditions, such as mass of fish is packed closely together in a pound or in a tank of chilling medium, favour the multiplication of anaerobic bacteria. It is characterized by the rapid development in localized parts of the individual fish of an obnoxious rotten egg like odour.

Fish so affected are commonly called as stinkers or bilgy fish. Microorganisms also affect the appearance and physical properties of several components of the body. The slime or skin and gills initially watery and clear, becomes cloudy, clotted and discoloured. The skin loses its bright iridescent appearance, bloom and smooth feel and becomes dull, bleached and rough to the touch. The peritoneum becomes dull and can be progressively more easily detached from the internal body wall. Microorganisms are the most important agents of deterioration in raw wet fish, since they give rise to the particularly undesirable flavours associated with spoilage. Thus, the control of deterioration is largely the control of microorganisms.

2. Non-microbiological deteriorations

They are further classified into enzymatic and non- enzymatic deteriorations.

a. Enzymatic deteriorations

Large numbers of different enzymes naturally present in the fish flesh are engaged in normal processes such as tissue building and muscular contraction and relaxation when fish are alive. But on death, these enzymes are involved in predominantly degradative reactions. One of these reactions is the gradual hydrolysis of glycogen to lactic acid during the first few hours resulting in a fall in pH from about 7.0 to 6.0 – 6.8 depending on species and the condition of the fish when the process is complete. The decline in pH is accompanied by the natural post-mortem stiffening called 'rigor mortis'. It affects quality in so far as the texture of the flesh is rendered somewhat firmer and its tendency to lose moisture when pressed. In terms of microbiological spoilage, it is an advantage to have the pH as low as possible for as long as possible. In practice, the generation of basic nitrogenous compounds like TMA and ammonia by microbiological action gradually raises the pH during the period after rigor mortis has passed off. In stale or putrid fish, the pH attains about 7.5 or even up to 8.0 in some species. The phenomenon of rigor just referred to is caused by another complex series of enzyme reactions lasting several hours or days depending upon temperature. Further reactions then render the flesh progressively more flaccid. The most significant enzyme deteriorations are those that affect flavour. The compounds responsible for the desirable sweetish, meaty and characteristic fish flavours of different species are changed by the intrinsic flesh enzymes to more neutral tasting compounds with the result that the fish become relatively more insipid. If this process of autolysis or self-digestion then continues sufficiently far it is believed that in many species the concentration of one particular breakdown product, hypoxanthine, becomes high enough to contribute to the bitter characteristics of unfresh fish.

The viscera or gut also contains enzymes, which is responsible for digesting food, when fish are alive. But on death, these powerfully digestive proteolytic enzymes attack the organs themselves and the surrounding tissues. The rate of attack is particularly great in fish that have been feeding heavily or feedy fish. In such fish, the organs quite quickly become degraded to a soupy structureless mass and the belly walls either digested away completely or so weakened that the slightest abrasion or pressure causes them to tear. This condition, known as ' belly burst or belly burn' is seen most often in pelagic species. Visceral enzymes are also capable of penetrating into the flesh and causing additional quality damage. The digestive enzymes of some shellfish are very active and are able to attack the flesh of even moribund fish. Because of these, such animals should be kept alive in as full vigour as possible until just before processing if the best quality product is to be obtained. Certain enzymes in the kidney tissue can dissociate trimethylamine oxide present in fish flesh into formaldehyde and dimethylamine. The enzyme, tyrosinase, particularly in the tail portions of crustaceans, cause the formation of black pigments called melanin in shrimps and lobsters, which affect their appearance.

b. *Non-enzymatic deteriorations*

The most prominent of the non-enzymatic deteriorations is the development of rancidity. This is caused by the attack of oxygen on the chemically unsaturated fatty acids in lipids contained in the fish flesh and other tissues. Fish in general have lipids of higher degree of unsaturation than most other foods and are therefore particularly prone to oxidative rancidity. The deterioration takes the form of the development of a linseed oil-like, painty odour and flavour in lean fishes and rancid off-flavours in fatty fishes. All fish tissues slowly loss fluid during storage and the amount of which varies with conditions, but may amount to 5 – 10% of body weight after 10 days in melting ice. This fluid or drip carries away some of the flavour compounds and so contributes to the general reduction in flavour. The effect is enhanced by leaching when the fish are constantly re-iced to compensate the ice melt water. Loss of weight through drip and loss of flavour through leaching are the quality loss in chill stored fish.

In some cases, it is impossible to attribute important quality deteriorations clearly to microbiological, enzymatic or non-enzymatic causes. Thus, during spoilage the blood contained in the kidney, its associated vessel and main artery along the backbone gradually diffuses into the adjacent flesh and causes discolouration and the exact mechanism of blood release is not known and although likely to be autolytic in nature. The appearance of the eyes of bony fish is a good guide to degree of spoilage. Very fresh fish have bright, convex eyes with a dark pupil; as spoilage progresses, the eyes become duller and grayer and pass from being flat to concave. Fresh fish flesh is translucent; stale fish tends to be opaque. With one exception, spoilage in raw fish is not associated with public health hazard. The exception is a type of poisoning called as histamine or scombroid poisoning. Naturally occurring spoilage bacteria such as *Proteus morganii* and some *Pseudomonas* strain act on the large amounts of histidine present in the fish flesh to produce biologically active histamine that when ingested in sufficient quantities give rise to a rarely fatal allergic type reaction. It is caused by eating spoiled meat from certain dark fleshed fish species such as mackerel, tuna and saury.

CHAPTER 3

───●───

HAZARDS IN AQUATIC FOOD

Hazard is a biological, chemical or physical agent in, or condition of, food with a potential to cause an adverse effect on the health of consumers. It is classified as biological, chemical and physical hazards. Biological hazards include the pathogenic bacteria, biogenic amines, viruses, parasites and aquatic biotoxins. Chemical hazards include the chemical contaminants with some potential for toxicity to human beings. Physical hazards include any potentially harmful extraneous matter not normally found in food.

3.1. BIOLOGICAL HAZARDS

The occurrence of pathogenic microorganisms in seafood is considered as biological hazards. They mainly include *Listeria monocytogenes, Clostridium botulinum, Salmonella, Vibrio cholerae* and other pathogenic vibrios. As per the National and International standards on aquatic food, these microbial pathogens should be absent in the products.

3.1.1. *Listeria monocytogenes*

Listeria monocytogenes causes the disease listeriosis and is of great concern to special risk groups. The susceptible groups include pregnant women and their fetuses, cancer patients and others undergoing immunosuppressive therapy, as well as diabetics and cirrhotics and the elderly. Although the risk of contracting listeriosis is less for normal, healthy individuals, they may also contract the disease. Typical *Listeria* infections result in septicemia, meningitis and encephalitis, although enteritis has been reported. Mortality is high among those infected as evidenced by a 29% death rate among patients in a New England outbreak associated with fluid milk. *Listeria monocytogenes* is an interesting pathogen because it is a facultative intracellular parasite. The organism enters the body through the intestine and has a variable incubation period that may be as short as 1 day to a month or longer. The ingested cells enter the body through ileal villi cells and are subsequently taken up by macrophage cells in the bloodstream. Instead of being digested, the engulfed cells multiply inside the host cell until the macrophage bursts and liberates the *L. monocytogenes* cells to repeat the process. This causes the transitory flu-like symptoms often reported at the initial stage of disease. The enteric phase of the disease is not consistent; some report upset stomach and diarrhea whereas other victims do not display these symptoms. The actual disease known as listeriosis does not occur until a severe form of septicemia, encephalitis, lesions, or meningitis develops. All of these forms of listeriosis may accompany infection of those who are not immunocompetent.

It has long been known that *Listeria* is widely distributed in the environment, humans and a variety of animals including seagulls. Most of these environmental strains are probably non-pathogenic. Other *Listeria* sp. than *L. monocytogenes* appears to be more common in tropical areas. There is little known association between seafood and listeriosis. An outbreak of perinatal listeriosis was reported from three obstetric hospitals in Auckland, New Zealand. Most of the 22 cases were due to strain 1b. The cause of the outbreak was not discovered but consumption of raw shellfish and raw fish may have played a role. Food surveillance for *L. monocytogenes* has increased since 1987 after the organism was found in refrigerated and frozen crabmeat. *Listeria* are relatively heat resistant but usually are not present in foods receiving an adequate heat treatment. The presence of *L. monocytogenes* in cooked seafood indicates cross-contamination of the product or under processing. *Listeria* are aerobic under most circumstances but can be facultatively anaerobic and thus grow well under reduced levels of oxygen in packaged products. *Listeria* can also survive and grow under adequate refrigeration temperature. The bacteria also survive freezing well, thus making adequate cooking and prevention of recontamination extremely important. Dairy products, salads and vegetables have been implicated in outbreaks of listeriosis. Contaminated food is increasingly recognized as an important vehicle of *L. monocytogenes*. Frequent isolations from seafood and the demonstration of growth potential in chilled smoked salmon at +4°C are evidence that seafood may be important in the transmission of *Listeria monocytogenes*.

Listeria are commonly identified by serotyping. Types 1-7 are known, with Types 1/a, 1/b, and 4b predominating as both environmental and clinical isolates. The serotyping scheme is based on both somatic and flagellar antigens. Phage typing has also been employed as a method of further identifying isolated strains. There are currently 27 phages used in the typing system of Audurier. Six species of *Listeria* are currently recognized, but only three species, *L. monocytogenes*, *L. ivanovii* and *L seeligeri* are associated with disease in humans and / or animals. However, human cases involving *L. ivanovii* and *L. seeligeri* are extremely rare with only four reported cases. *L. monocytogenes* is subdivided into 13 serovars on the basis of somatic (O) and flagellar (H) antigens. This subdivision is of limited value in epidemiological studies since most of the isolates belong to three serotypes. More valuable methods are phage-typing, isoenzyme-typing, and DNA-fingerprinting.

Currently FDA in the US requires that *L. monocytogenes* be absent in ready-to-eat seafood products such as crabmeat or smoked fish. This restriction does not apply to raw product that will be cooked before eating. Other countries have similar regulations, which are completely unrealistic. Due to the ubiquitous nature of *L. monocytogenes,* such products cannot be guaranteed free of *L. monocytogenes*. The FDA is now considering possible changes in their policy. Products will be classified according to known, established risks. A zero-tolerance will still be maintained for products, which have received a listericidal treatment, as well as for products, which have been directly implicated in a food borne outbreak. Low number of *L. monocytogenes* may then be allowed in other types of products, particularly those in which the organism can be shown to die-off. There is a general agreement between microbiologists that the presence of low numbers of *L. monocytogenes* in food may need to be tolerated. They suggest that a limit of 100 *L. monocytogenes* is reasonable. But, they feel that greater than 10 *L.*

monocytogenes / g is likely to constitute a risk to human beings particularly predisposed persons (very old, very young or immuno-suppressed). These quoted figures should be compared to the background level of *L. monocytogenes* in foods, which is approximately 1-10 *L. monocytogenes* /g. This means that little or no growth of *L. monocytogenes* in foods should be tolerated. Proper GMP and factory hygiene can maintain the level of *L. monocytogenes* contamination on fish products at very low level of less than 1-10/g. Chlorine based, iodine-based, acid anionic, and quaternary ammonium-type sanitizers are effective against *L. monocytogenes* at concentrations of 100 ppm, 25-45 ppm, 200 ppm and 100-200 ppm, respectively. Further disease control with products, which have not been subject to listericidal processing, rest with control of growth of this organism in the products. It will be noted that is difficult to control in chilled fish products such as cold smoked fish. The organism can grow at temperature down to +1°C, and it is tolerant to NaCl up to 10% at neutral pH and 25°C. Nitrite is not inhibitory to *L. monocytogenes* at permitted levels unless there is an interaction with other inhibiting agents. Listericidal processing consists primarily of heat treatment. The heat resistance of *L. monocytogenes* has been the subject of extensive investigation particularly for milk and dairy products. The thermal death time curve (TDTC) for *L. monocytogenes* in cod and salmon was studied. The results show a significantly higher heat resistance of *L. monocytogenes* in salmon fillets compared to cod fillets with D_{60} being 4.5 min. in salmon and 1.8 min. in cod.

3.1.2. *Clostridium botulinum*

Clostridium botulinum is classified into seven toxin types (A to G). The types (A, B, E and F) pathogenic to humans can be conveniently divided into two groups, viz. proteolytic types A, B and F, which are heat resistant, mesophilic, NaCl-tolerant and have the general environment as the natural habitat and non-proteolytic types B, E and F, which are heat sensitive, psychrotolerant, NaCl-sensitive and have the aquatic environment as the natural habitat. Toxins produced by *C. botulinum* types A, B, E and F are the cause of human botulism. The disease can vary from a mild illness to a serious disease, which may be fatal within 24 h. In most cases, the symptoms develop within 12 to 36 h. These are generally nausea and vomiting followed by neurological symptoms such as visual impairment, loss of normal mouth and throat function, lack of muscle coordination and respiratory impairment, which is usually the immediate cause of death. Type E botulism tends to have most rapid onset of symptoms, while type A botulism tend to be the most severe.

Majority of botulism outbreaks in the northern and temperate regions are associated with fish, and in general type E was the responsible type. Types A and B botulism have been generally associated with meat or meat products, but fish and fish products have also been vehicle for those types. All types of fish products except raw fish to be cooked immediately before consumption have been involved in outbreaks of botulism, but the majority of outbreaks have been associated with fermented fish. Botulinal toxin is one of the most potent of all poisons, and the amount needed to cause death in humans has been estimated to be as low as 30 - 100 ng. The toxin is sensitive to heat and pH above 7. For safe inactivation of any botulinal toxin at concentrations up to 10^5 LD/g in foods, time / temperature combinations of 20 min. at 79°C or 5 min. at 85°C have been recommended. Normal household cooking and frying of raw fish products are therefore sufficient to destroy any pre-formed toxin. While botulinal toxin is rapidly destroyed in fish products with pH >7.5, it is extremely stable in a salty and acid environment. Hence, botulinal toxin formed in the raw material will be found again or even increased in-situ in the final products such as heavily salted, marinated or fermented fish.

Spores of the non-proteolytic *C. botulinum* types, particularly type E are widely distributed in the aquatic environment in the temperate and arctic zones. Thus, up to 100% of sediment samples from coastal areas, particularly in closed waters and from aquaculture ponds may contain this organism. Distribution patterns of *C. botulinum* type E suggest that this is a true aquatic organism and that multiplication occurs in-situ. A much lower prevalence is found in live fish although upto 100% of fish from aquaculture and coastal waters may carry this organism. Fish caught in the high seas are generally free from *C. botulinum*. In warm tropical waters and in fish from these areas, other types than type E are frequently found. Proteolytic *C. botulinum* is frequently found in soil and the terrestrial environment. Animals, both vertebrates and invertebrates, have an important role in both the distribution and build up of botulinum spores. Spread of spores from the terrestrial environment to the aquatic environment including the fish in these areas is therefore a distinct possibility as well as spread of spores to the fish processing environment. Being mesophilic, the proteolytic types do not have the same possibilities for multiplication in nature as type E.

Main factors that control growth of *C. botulinum* in foods are temperature, pH, a_w, salt, Eh and added preservatives. The natural level of *C. botulinum* in fish is much lower than levels used in most challenge studies. Initiating growth and toxin production will therefore be much delayed at comparable conditions. The associated flora in fish products may cause spoilage before the product becomes toxic. Some microorganisms may also inhibit *C. botulinum*. It is an anaerobic organism preferring a low Eh for growth. The Eh for fish and fish products is high and this may cause delay in growth and toxin production at otherwise comparable condition in bacteriological media. The presence of an associated or spoilage microflora may, however, also add to the risk, as this microflora may use oxygen and facilitate the growth and toxin production by *C. botulinum* type E.

A combination of salt and low temperature very effectively inhibits toxin production and growth of *C. botulinum*. D-value varies considerably among *C. botulinum* types and even among strains within the same type. The spores of the non-proteolytic types are considerably less resistant than the proteolytic types. The heat resistance of non-proteolytic types is particularly important for mildly heat-treated, pasteurized products, where conditions for growth are excellent for surviving spores. D-values at 82°C for these products may vary from 0.5 to 2 min. A minimum heat treatment of 90°C for 10 min. should provide a safety factor of 10^6 (a 6-D process or a 6-log reduction of spore count) for non-proteolytic *C. botulinum*. Spores from proteolytic *C. botulinum* are much more heat resistant. In general, D121 values are in the range of 0.1 - 0.25 min. These spores are a particular concern in the sterilization of low acid canned foods, and the canning industry has adopted a D-value of 0.2 min. at 121°C as a standard for calculating thermal processes. For the most resistant strains, Z-values are approximately 10°C, which has also been adopted as a standard. Control of *C. botulinum* in fishery products can be achieved by inactivation of spores or by inhibition of growth. Current guidelines regarding safety with respect to

C. botulinum includes one of the following procedures such as storage at all times at < 3.3°C, storage at 5 - 10°C and a shelf life of < 5 days, a heat treatment of 90°C for 10 min. combined with chill storage (< 10°C), a pH of 5.0 throughout the food combined with chilled storage (< 10°C), a salt-on-water concentration of 3.5% or a_w of 0.97 throughout the food combined with chill storage (< 10°C). It should be noted that products where growth of non-proteolytic *C. botulinum* is completely inhibited by salt or low pH or inactivated still has a requirement for chilled storage. The reason is that the proteolytic *C. botulinum* may still be able to grow if temperatures are > 10°C. The US requirement states that vacuum-packed cold smoked fish contain 3.5% NaCl or 3% if combined with 100 - 200 ppm nitrite. For air-packaged fish, not less than 2.5% NaCl in the muscle is required.

The canning industry has adopted a 12-D process as a minimum heat process applied to commercial canned low acid foods. The heat required to provide this "botulinum cook" or a 12-decimal reduction in proteolytic *C. botulinum* spores is therefore equal to 12 x D_{121}-value or 12 x 0.2 = 2.4 min. at 121°C. The highest known D_{121} value is 0.25 min., which gives a F- value of 12 x 0.25 = 5. Using F- values between 2.4 -3 has led to safe production of canned low acid food for many decades. Often higher F values are used in commercial practice. Refrigeration is often regarded as the primary method of preservation of fresh foods, including seafood. At temperatures below 10°C, there is no risk of toxin production by proteolytic *C. botulinum* types A and B. At higher storage temperature, additional preservation or treatment is required to produce safe food.

3.1.3. *Salmonella*

The occurrence of *Salmonella* in fish and shellfish, either in fresh or marine waters, has normally been associated with fecal contamination of the area from which they were harvested. Concern over *Salmonella* contaminated seafood is not new. In the early years, the shellfish industry was plagued with *S. typhi* as a primary pathogen found in raw shellfish harvested from polluted waters. Reports as early as 1954 have indicated that *Salmonella* could be also consistently isolated from finfish harvested in highly polluted waters. Survival in water depends on many parameters such as biological (interaction with other bacteria) and physical factors (temperature). It has been demonstrated that both *E. coli* and *Salmonella* sp. can multiply and survive in the estuarine environment for weeks, while presented similar results on survival in tropical freshwater environments. There is some indication that both freshwater and marine fish exposed to *Salmonella* in polluted waters remain positive for up to 30 days after exposure. The salmonellae are rod-shaped, non-spore-forming, mesophilic, Gram–negative bacteria belonging to the family Enterobacteriaceae. There are more than 2,500 serotypes of Salmonellae recognized at this time and more are added to the list every year. The species name for *Salmonella* is determined by unique serotype determination of somatic and flagellar antigens. The primary location of *Salmonella* spp. is the alimentary tract of mammals, birds, amphibians, and reptiles. Salmonellae are not endemic to the intestinal tracts of finfish, crustaceans, or molluscs. Because this organism is faecal in origin, it can be found in processed seafood through pollution of the water environment or by contamination of the seafood after catch.

The most common disease associated with *Salmonella* is gastroenteritis, commonly referred to as salmonellosis. Symptoms of acute salmonellosis are nausea, vomiting, abdominal cramps, diarrhea, fever, and headache. The onset time is usually 6-8 h but may be no longer with low doses of virulent strains. The infective dose for *Salmonella*, which is quite variable and dependent on serotype and other factors, may range from only a few organisms to over 10^5 to cause illness. The general susceptibility of the person exposed is as important as other factors in illness. The infectious does may be as low as 15-20 cells, depending on varying virulence factors, age and general health status of the victim. Persons of all ages are susceptible to *Salmonella* infections but symptoms are more severe and prolonged among the elderly, infants, and people with underlying illnesses. The acute symptoms may be short in duration (1-2 days) or extended (1 week or more). The duration of illness seems to be affected by the same factors as infectious dose. Several complications can arise from acute salmonellosis. These are septicemia, local lesions, and reactive or sterile site arthritic conditions. The most serious forms of infection are typhoid and paratyphoid fever caused by *S. typhi* and the *S. paratyphi*. The septicemia caused by these organisms can be life threatening if not successfully treated. The septicemia and organ lesions caused by *S.*

typhi and *S. paratyphi* A, B, and C are the cause of prolonged clinical illness of great severity. The fatality rate for typhoid fever is 10%. There are other species that have a percentage of human septicemia associated with infection. *Salmonella enteritidis* has shown a 3.6% mortality rate among hospital and nursing home outbreaks. The presence of any type of *Salmonella* contamination of food should be taken seriously because of these grave consequences to enteritis. *Salmonella* septicemia has been associated with subsequent infection of all body organs and a severe arthritis has been reported as associated with a small percentage of food borne disease victims. In addition to reactive arthritis, Reiter's syndrome, sometimes referred to as Reiter's triad, occurs. In this situation joint inflammation is accompanied by conjunctivitis and urethritis.

Aquatic food can be contaminated with *Salmonella* in aquaculture systems from many sources including farm runoff and direct fecal contamination from livestock or feed. As an example, domestically produced, pond reared catfish carry *Salmonella* at an apparently low level of contamination. The flow of *Salmonella* in the food chain can be complex. This is especially true in aquaculture systems. The role of fish meal, for instance, can ultimately affect the food supply of other commodities. *Salmonella* can survive drying and thus be a problem in fish meal that is fed to livestock and poultry. Concern for the prevalence of *S. agona* in fish meal became a national concern in the 1960s. Smoked fish have been a traditional source of salmonellosis outbreaks and minor epidemics. Typical of these outbreaks was an instance involving *S. newport* where the cause of three separate outbreaks traced to one fish processing plant that has many elements enabled the outbreak to occur. Proper hygiene was not observed and the fish were temperature abused subsequent to their sale. The plant was processing fish under extremely poor sanitary conditions. A human carrier who packed the smoked fish and had been previously ill with salmonellosis was identified and product temperature abuse also contributed to this series of outbreaks. A second epidemic from *S. java* contaminated white fish illustrates similar points.

Salmonella montevideo is highly pathogenic to human and causes gastro-intestinal diseases. The first outbreak of *S. montevideo* infection was reported in Australia and Newzealand during 2002.

S. montevideo has been isolated predominantly from fish meal and ready-to-use compound fish feeds used in aquafarms and the main source is presumed to be from feeds or any fertilizers prepared using poultry or cattle. *S. montevideo* strains have also been isolated from shrimp samples collected from fresh food markets during 2010.

Molluscan shellfish species require special sanitary practices from harvest to consumption because they are frequently consumed raw. The early public health safeguards put in place for shellfish were developed to protect the public from outbreaks of typhoid fever from oysters and clams infected with *S. typhi*. In the case of *S. typhi* infection, the source of infection is either from sewage pollution or a carrier of the organism. *Salmonella* is occurring commonly in domestic animals and birds and many are asymptomatic *Salmonella*-excreters. Raw meat and poultry are therefore often contaminated with this organism. Most literature reports indicate that seafood is a much less common vehicle for *Salmonella* than other foods, and fish and shellfish are responsible for only a small proportion of total number of *Salmonella* cases reported in the World. Most prawns and shrimps are cooked prior to consumption and the products therefore pose minimal health risks to the consumer except by cross contamination in kitchens.

3.1.4. *Vibrio cholerae and* other pathogenic Vibrios

Vibrios are Gram-negative, non-spore forming, facultatively anaerobic, motile and straight or curved rods belonging to the family Vibrionaceae. All species except two are oxidase-positive. With few exceptions, they produce acid but no gas from glucose. Most are halophilic in nature. They became prominence even during the 19th century due to the cholera outbreak caused by *Vibrio cholerae*. Seven pandemics of cholera have so for been attributed to *V. cholerae* O1 serovar. The severe outbreaks of cholera–like illness in India and several other countries due to a new serotype designated as *V. cholerae* O139 indicates the eighth pandemic of cholera. Vibrios are widely distributed in the marine environment. *Vibrio* infections are acquired either through contact with marine environment or through consumption of seafood in humans. The genus comprises 34 species of which 13 species can cause human disease, including wound infections, septicemia and gastroenteritis. Seafood-borne diseases are primarily caused by *V. cholerae, V. parahaemolyticus* and *V. vulnificus. V. cholerae* and *V. parahaemolyticus* cause gastrointestinal disease, while *V. vulnificus* causes septicemia. Pathogenic vibrios have been found to produce a variety of extracellular products such as enterotoxins, cytolysins, cytotoxins, hemolysins, proteases, collagenases, mucinases and siderophores that play vital role in fluid accumulation, cell lysis and tissue destruction. Of which, only cholera toxin (CT) produced by *V. cholerae* O1 has been conclusively known to be associated with distinct gastrointestinal syndrome. Some strains of *V. cholerae* non–O1 and *V. mimicus* also produce CT indicating their potential to cause cholera–like illness in human.

V. cholerae is serologically divided into two groups, serotype O1 and non-O1. It can also be divided into two biovars viz. ElTor and Classical. The toxigenic strains are capable of producing cholera toxin or a very similar toxin. It is actively motile by a single polar flagellum. On microscopic observation, it moves faster and often goes out of field in a zigzag manner. This motility known as gum-shot motility or darting motility is used as an important characteristic for identification. It grows well at 37°C and is tolerant to alkali, grows in a pH of up to 9.2. It is widely distributed, occurring in environments where there is no human pollution, and considered to be indigenous to many estuaries and other aquatic environments. It is capable of surviving for a long period of time in both the freshwater and seawater. It is capable of entering into a viable but nonculturable (VBNC) phase during adverse conditions such as low temperature and lack of nutrient availability.

V. cholerae infections are through oral route, by contaminated water or food. It attaches to the intestinal mucosa and produces enterotoxin that causes disruption in the cyclic AMP system resulting in water and electrolyte loss. Clinical symptoms vary from the presence of mild, watery diarrhea to acute diarrhea (known as rich water stool), abdominal cramps, nausea, vomiting, dehydration, shock and may lead to death. *V. cholerae* non-O1 infections are associated with exposure to natural aquatic environments, consumption of seafoods, exposure to polluted waters. Strains of non-O1 *V. cholerae* appear more capable than *V. cholerae* O1 for survival and multiplication in a wide range of foods. Environmental hygiene and sanitation play a significant role in the control of contamination of seafoods with *V. cholerae*. Processing plant surroundings should be kept clean and disinfected. Personnel hygiene of fish handlers, proper chlorination and use of potable water in processing plants are important in the control of *V. cholerae* in seafoods.

Vibrio parahaemolyticus is yet another organism causing food poisoning associated with the consumption of raw fish. This was first isolated in Japan in early 1950s. It grows faster as it has a generation time 7 min. Hence, this bacteria multiplies rapidly in fish that has been contaminated with this organism and subject to temperature abuse during storage causes food poisoning. It grows between temperatures of 15°C to 43°C, prefers alkaline conditions (pH 7.6 – 8.6) and a salt concentration of 2-4% NaCl. It is sensitive to chilling and freezing. Heating to 100°C completely destroys this organism. It can be divided into two types depending on its haemolytic property. The strains, which are haemolytic, are called as Kanagawa-positive and those that are non-haemolytic are Kanagawa-negative. Studies have shown that about 97% of the cultures isolated from human infections were Kanagawa-positive. The food poisoning caused by *Vibrio parahaemolyticus* is in the form of gastroenteritis in most cases. Symptoms include abdominal cramps, diarrhea, nausea and vomiting without bloody stools, although a dysenteric illness may develop. The incubation period ranges from 4 to 48 h and lasts approximately 72 h. Microbiological limits for fresh and frozen seafood with reference to vibrios by several importing countries and regulatory agencies have given a zero tolerance for *V. cholerae*. However, an acceptable limit of 100/g for *V. parahaemolyticus*, with 1000/g as a boundary between marginal acceptance and unacceptability has been specified.

V. vulnifus is an organism of great concern in aquatic food safety due to the severity of the disease and the high mortality rate it can cause. In addition, *V. vulnificus* is a potentially lethal food-borne pathogen and capable of causing primary septicemia and necrotizing wound infections in susceptible individuals. The virulence factors associated with *V. vulnificus* include a capsule, cytolysin, protease/ elastase and phospholipase, but these are found in nearly all clinical and environmental strains. *V. vulnificus* is the leading cause of death from seafood in the United States, with approximately 40 cases each year. Outbreaks of *V. vulnificus* have also been reported in Europe and Asia. According to the CDC's most recent data, there were more than 900 reported cases of vibriosis between 1998 and 2006 in the Gulf Coast region, including Alabama, Florida, Louisiana, Texas and Mississippi. About 38.5% of clams obtained in 2017 from Mangalore fish market, South-West coast of India have been shown to harbour *V. vulnificus*. Few seafood industries from Tamil Nadu coastal region faced the recent rejection of seafood in South Africa and other countries due to the presence of *V. vulnificus*. It is usually found worldwide in coastal or estuarine environments with water temperatures from 9 to 31°C. The organisms preferred habitat is considerably more selective and has been reported to be water temperatures in excess of 18°C and salinities between 15 to 25 ppt. *Vibrio vulnificus* infection is transmitted by eating contaminated seafood or by exposure to seawater through an open wound. The ability of *V. vulnificus* to cause disease is associated with the production of "Multifunctional-Auto processing RTX" (MARTXVv) toxin that encodes by RtxA1 gene. HlyU gene regulates the expression of the repeat-in-toxin (RtxA1) gene. The hemolysin *vvhA* gene plays an additive role for pathogenesis of *V. vulnificus*.

3.2. CHEMICAL HAZARDS

The chemical hazards in aquatic foods are divided into three major groups. They are naturally occurring chemicals, unintentionally or incidentally added chemicals and food additives. Naturally occurring chemicals are histamine poisoning, poisoning due to biotoxins such as tetradontoxin, ciguatoxin and shellfish toxins like paralytic shellfish poisoning (PSP), diarrhetic shellfish poisoning (DSP), amnesic shellfish poisoning (ASP) and neuorotoxic shellfish poisoning (NSP).

3.2.1. Histamine poisoning

Histamine poisoning or scombroid poisoning is caused by high level of histamine in the muscle of fish. It is mostly associated with scombroid fishes and other dark muscle fishes. About 50 species including some popular fishes like tuna, mackerel, blue fish, dolphin, carangid, herring, sardine and anchovy have been found to have a potential threat of histamine poisoning. They contain higher concentration of histidine. On spoilage it is converted to histamine by bacterial groups like *Morganella morganii, Klebsiella pneumoniae* and *Hafnia alvei.* Most of these bacteria are found in fish as a result of post-harvest contamination. They grow well at 10°C. Large amount of histamine is formed by *M. morganii* at low temperature (0-5°C) followed by storage up to 24 h at high temperature (10-25°C). Histamine production increases with temperature and 37°C is the optimum temperature. Low temperature storage from capture to storage is the most effective control measure. Storage over 0°C or very near to 0°C limits histamine formation. Fluctuations in storage temperature may result in histamine production. Histamine is resistant to heat. So if the fish is cooked, canned or heat treated it is not destroyed. The human body tolerates 50 mg of histamine without any reaction. The maximum permissible level of histamine in fresh seafood is 20 mg/100g.

Incubation period is very short from few minutes to few hours. The most common symptoms are facial flushing, edema and gastrointestinal disturbances such as nausea, vomitting and diarrhea. Neurological symptoms like headache and burning sensation in the month are also common. The duration of the period of illness is less and usually few hours.

3.2.2. Biotoxins

Marine biotoxins cause a number of seafood borne diseases. Two main type of poisoning due to seafood is recognized. They are fish poisoning caused by eating fish containing poisonous tissues, and shellfish poisoning resulting from ingestion of shellfish that have concentrated toxins from the plankton. The sources of biotoxins and the organs involved in accumulation of biotoxins is given in Table 1.

Table 1. Some of biotoxins and organs involved in bioaccumulation

Toxin	Source	Organs involved
Tetrodotoxin	*Vibrio alginolyticus*	Ovaries, liver and intestine of puffer fish
Ciguatera	Marine algae (*Gambierdiscus toxicus*)	Muscle of coral reef fish

[Table Contd.

Contd. Table]

Toxin	Source	Organs involved
Paralytic shellfish poisoning (PSP) (Saxitoxins)	Marine dinoflagellates *(Gonyaulax catenella, Pyrodinium* spp.)	Digestive glands and gonads of molluscs
Diarrhetic shellfish poisoning (DSP) (Okadic acid)	Marine dinoflagellates (*Dinophysis* spp.)	Digestive glands and gonads of molluscs
Neurotoxic shellfish poisoning (NSP)	Marine dinoflagellates (*Gymnodinium breve*)	Digestive glands and gonads of molluscs
Amnesic shellfish poisoning (ASP) (Domoic acid)	Marine diatoms (*Nitzschia pungens*)	Glands of molluscs

3.2.2.1. *Tetradotoxin*

This toxin is the most lethal of all the poisons. It is produced by *Vibrio alginolyticus* found in certain algae and stored in the liver, ovaries and intestine in various species of puffer fish. Puffer fish is a delicacy in Japan. This poisoning can also be called as puffer fish poisoning. The muscle is free of toxin. Tetradotoxin causes neurological symptoms that develop 10-45 minutes after ingestion. Symptoms such as gastrointestinal, vomitting, diarrhea, neurological tingling sensation in face and extremities, paralysis, respiratory failure and cardiovascular collapses are encountered.

3.2.2.2. *Ciguatera*

Ciguatera is the most common form of fish poisoning. It is caused by the ingestion of fish that have become toxic by feeding on toxic marine algae. The toxic agent originates from a blue green algae (*Gambierdiscus toxicus*) which is the passed onto herbivorous fish and indirectly to carnivorous fish. More than 400 tropical and subtropical food fishes are involved in this poisoning. Fish species such as red snapper (*Lutjanus bohar*), grouper (*Variola louti*) and moray eel have been reported to contain this toxin. Among the organs, liver is more toxic followed by the viscera and muscle. Symptoms include gastrointestinal and neurological disorders like vomiting, diarrhea, tingling sensation, ataxia weakness. Duration of illness may be 2-3 days, but sometimes it extends for more days. Death results from circulatory collapse.

3.2.2.3. *Paralytic shellfish poisoning (PSP)*

PSP has been associated with dinoflagellate (*Gonyaulax catenella* and *Pyrodinium* spp) blooms (>10^6 cells/liter), which may cause discolouration of water. Mussels, clams, oysters, scallops, which feed on these toxic dinoflagellates, retain the toxin in siphon and become toxic. The toxin is known as saxitoxin and it is stable to heat and hence not destroyed by cooking. It is highly toxic; 1 g of toxic shellfish meat can kill 5 people. Estimated lethal dose for human is 1 to 4 mg. Symptoms of poisoning include tingling, burning and numbness of lips and fingertip, drowsiness etc. The toxin depresses respiratory and cardiovascular regulating centers and death usually occurs from respiratory failure. After a number of mouse bioassay tests, the US FDA has set limits for this toxin in shellfish meat marketed for human consumption as 80 mg/100g. A mount unit (MU) is defined as the minimum amount of poison that would kill a 20 g mouse in 15 min. when the toxin is injected in 1 ml of extract intraperitoneally.

3.2.2.4. *Diarrhetic shellfish poisoning (DSP)*

DSP normally accumulates in shellfish from the dinoflagellates, *Dinophysis fortii*. The most widely distributed toxin is okadic acid and its derivatives. The primary symptom is acute diarrhea with vomiting and abdominal pain and the victims recover within 3-4 days. No fatalities have ever been recorded.

3.2.2.5. *Neurotoxic shellfish poisoning (NSP)*

Neurotoxic shellfish poisoning has been described in people who consume bivalves that have been exposed to red tides caused by dinoflagellates (*Gymnodinium breve*). This disease has been limited to the Gulf of Mexico.

3.2.2.6. *Amnesic shellfish poisoning (ASP)*

Amnesic shellfish poisoning (ASP) has only recently been identified. It is due to domoic acid, an amino acid produced by the diatom *Nitzschia pungens*. The symptoms of ASP vary greatly from slight nausea and vomiting to loss of balance. The central nervous system is affected resulting in confusion and memory loss. The memory loss may be permanent in surviving patients and hence the name "amnesic shellfish poisoning". The control of marine biotoxin is difficult and disease cannot be entirely prevented. Only by estimation, the presence of these toxins may be assessed.

Detection of biotoxins

Mouse bioassay or HPLC are the two important methods for detecting biotoxins (Table 2).

Table 2. Methods of detecting biotoxins

Biotoxins	Level of tolerance	Method of analysis
Ciguatera	Control not possible	No reliable method
PSP	8.0 mg/100 g	Mouse bioassay, HPLC
DSP	0-60 mg/100 g	Mouse bioassay, HPLC
NSP	Any detectable level / 100 g is unsafe	Mouse bioassay, no chemical method
ASP	20 mg/ g Domoic acid	HPLC

3.2.3 Antibiotic and Pesticide residues

The most common toxic residues encountered in fish and fishery products are classified as metallic residues and pesticide residues. With the development of aquaculture industry, yet another element of risk came into reality i.e. antibiotic residues.

3.2.3.1. *Antibiotic residues*

Antibiotics are substances of natural, semi-synthetic or synthetic origin that exhibit antibacterial activity. Antibiotic drugs are being used in aquaculture to treat and prevent disease; control parasites; and affect growth and reproduction. Drugs are also added to fish feed to improve the efficiency of their conversion into edible tissue. Unregulated / unapproved drugs administered to farmed fish pose a potential health hazard and environmental problems. These substances may be carcinogenic, allergenic and /or may cause antibiotic resistance and hormone imbalance to consumers.

Effects of antibiotic residues

1. *On environmental microflora*

Aquaculture environment is a very dynamic system. The mineralization process caused by microflora, mainly bacteria, maintains the chemical/ biochemical/ gaseous equilibrium in the farm. In shrimp farms, on an average, $60mg/m^3$ of waste is resulted for every kg of shrimp produced, which is mineralized by the soil/farm microorganisms. Many chemicals used in aquaculture degrade very rapidly, while some antibiotics like oxytetracycline, oxolinic acid and fluoroquinolones persist for at least 6 months. These residues destroy the environmental microflora and delay the natural mineralization process. So, the waste accumulates in the farm and causes favourable environment for the disease. They also affect the non-target and non-cultured fish in the adjacent waters due to secondary medication.

2. *Development of drug resistant bacteria including human pathogens*

Excessive use of antibiotics causes loss of efficiency of prophylactic antibacterial agents through the development of resistant strain. Antibiotic resistance results either from a mutation or is due to a plasmid (e.g. chromosomal addition through transformation, transduction, or sexual reproduction). Use of fluoroquinolones and glycopeptides increase drug resistant in pathogens and these pathogens get transferred to humans on consumption.

3. *Retention of drug residues in farmed shrimps*

An animal exposed to an antibiotic drug imbibes it into the body and their elimination depends on its chemical nature. The antibiotics used in aquaculture accumulate in shrimp tissue and exoskeleton. Some drugs are eliminated within hours of application, while some other residues remain as long as 200 days. The drug withdrawal period before harvesting (i.e. 2-3 weeks) for shrimp does not eliminate the accumulated residues from the tissues, particularly exoskeleton. The detection of antibiotic residues in Indian seafood has raised objection from importing nations. The maximum residual limit (MRL) has been introduced and is followed by food safety experts. The MRL is attained for each aquaculture species by imposing withdrawal period following treatment.

Permitted antibiotic residues

1. FDA approved aquaculture drugs

Relatively few drugs have been approved for aquaculture. This made fish farmers to use unapproved drugs that are not labeled for drug use and approved drugs in a manner that deviates from the labeled instructions. When a drug is approved by FDA's Center for Veterinary Medicine, the conditions of the approval are listed on the label of the product, which include the species of the drug approved for, dosage, route of administration, frequency and indications for use. FDA approved aquaculture drugs with their approved sources; species and withdrawal times are listed below (Fish and Fisheries Products Hazards and Controls Guidance: 3rd Edn, June 2001) in Table 3.

Table 3. FDA approved aquaculture drugs

Drug	Species	Withdrawal time	Tolerance limit
Oxytetracycline	Salmonid	21 d	2.0 ppm
	Catfish	21 d	2.0 ppm
	Lobster	30 d	2.0 ppm
Sulfamerazine	Trout	21 d	Nil
	Salmonid	42 d	ppm
Sulfadimethoxin/ Oremethoprim	Catfish	3 d	0.1 ppm

2. European Union Standards

The EU standards for aquaculture drugs (Annexure III to the EEC Regulations No. 2377/90) are given in Table 4.

Table 4. EU standards for aquaculture drugs

SI. No.	Antibiotics	MRL (in ppm)
1.	Antibiotic residues (Anti-infection agents, antibiotics and quinolones)	Nil
2.	Sarafloxacin (Salmonidae)	0.03
3.	Nafacillin (in bovine tissue)	0.3

3. *Indian Standards*

The Indian standards for aquaculture drugs are as per Government order dated 17.08.2001 of the Gazette of India (Extraordinary), Part II-Section 3, Sub-section (ii) No. 582 of Govt. of India, New Delhi, which are given in Table 5.

Table 5. Indian standards for aquaculture drugs

Sl. No.	Antibiotics	MRL (in ppm)
1.	Chloramphenicol	Nil
2.	Furazolidone	Nil
3.	Neomycin	Nil
4.	Tetracylcine	0.1
5.	Oxytetracycline	0.1
6.	Trimethroprim	0.05
7.	Oxolinic acid	0.3
8.	Nalidixic acid	Nil
9.	Sulphamethoxazole	Nil

Control measures

To control this hazard, all drugs, whether for direct medication or for addition to feed must be approved by FDA. Under certain conditions, unapproved drugs may be used in conformance with the terms of an Investigational New Animal Drug (INAD) application. Withdrawal time must be observed to ensure that the edible tissue is safe when it is offered for sale. Labels of approved drugs list mandatory withdrawal times. Control measures for approved aquacultue drugs used in aquaculture operations can include the following:

1. On-farm visits to review drug usage before receipt of the product
2. Receipt of supplier's lot-to-lot certification of proper drug usage
3. Review of drug usage records at receipt of the product
4. Drug residue testing
5. Receipt of evidence that the producer operates under a third party-audited quality assurance programme for aquaculture drug use

Methods for the determination of antibiotic residues

Drug residues analysis is an important aspect of quality control for fish processing industries. This poses special problem for the analyst, as animal products are highly complex materials containing protein, lipid and low molecular weight components. The methods used at present are classified into three groups:

I. Microbiological method

II. Electrophoretic method

III. Physicochemical method

I. Microbiological method

It is the most frequently used detection method. It exploits the sensitivity of certain bacterial strains to one or several antibiotics. The manifestation of this inhibition is affected in solid media (e.g. agar diffusion method). When one or more antibiotics in a solution are brought into contact with an agar medium they diffuse into it. The diffusion is proportional to the logarithm of their concentrations. The growth of the test microorganism cultured in agar after incubation shows the presence of an inhibiting substance through the appearance of a translucid zone in the antibiotic diffusion zone, while everywhere else the growth of the microorganism is visible. Agar diffusion method is relatively very rapid and does not require much laboratory equipment. It is especially well adapted when the sample antibiotic is known. The sensitivity of certain test bacteria to different antibiotics is given in Table 6.

Table 6. Sensitivities of certain test strains to various antibiotics

Antibiotics	Test microorgansims				
	E. coli	Sarcina lutea	Bacillus subtilis	Bacillus cereus	Bacillus megaterium
Tetracycline µg/ ml	0.16	0.08	0.08	0.08	0.08
Pencillin G µl/ml	2.50	0.003	0.005	2.50	0.005
Chlorotetracycline µg/ ml	0.31	0.16	0.08	0.08	0.08
Neomycin µg/ ml	2.50	0.08	0.04	0.62	0.02
Oxytetracycline µg/ ml	0.16	0.08	0.16	0.08	0.16

[Table Contd.

Contd. Table]

Antibiotics	Test microorgansims				
	E. coli	*Sarcina lutea*	*Bacillus subtilis*	*Bacillus cereus*	*Bacillus megaterium*
Nitrofurantoine µg/ ml	1.25	0.62	2.50	1.25	2.50
Erythromycin µg/ ml	0.62	0.62	0.04	0.04	0.02
Novobiocin µg/ ml	0.31	0.02	0.08	0.08	0.16
Bacitracin µg/ ml	1000	0.004	5.00	0.31	0.004

II. Electrophoretic method

This electrophoretic technique was first used in 1979 by associating it with the microbiological method. The sample to be tested is deposited in hollow cavities in the agar or in wells. Under the effect of an appropriate electric current, the antibiotics separate with different and specific speeds and directions of migration, which provides information on the nature of antibiotics present in the sample. A second layer of agar cultured with one or more test microorganisms detects the diffusion of the antibiotics into the gel after incubation. A positive detection is indicated by the formation of clear zones, which is proportional to the concentration of the antibiotic present. This technique is rarely used in routine work. It is an excellent reference method for situations requiring expertise.

III. Physicochemical method

a. Radioenzymatic method
b. Radioimmunological method
c. Immunoenzymatic method
d. Fluorimetry method
e. Bioluminescence method
f. Spectrometry method
g. Chromatographic methods

i. Thin layer chromatography

Antibiotic residues are identified on the plates by their retention factor. Determinations can also be affected by fluorescence. The detection limits are tetracycline – 0.025 µg/ ml; chloramphenicol – 1.0 µg/ ml; neomycin – 15 µg/ ml; streptomycin – 0.5 µg/ ml.

ii. Gas chromatography

Different protocols for extraction, identification and determinations of antibiotics by gas chromatography are available. The detection limits are tetracycline – 0.5 to 10µg/ ml; chloramphenicol – 0.01 µg/ ml; pencillin G– 0.005 UI/ ml.

iii. High performance liquid chromatography

This technique is widely used for the detection and determinations of most antibiotics. The extraction of antibiotics is followed by purification before separation by HPLC.

3.2.3.2. Pesticide residues

Pesticides are substances or preparations used to combat living beings harmful to humans excluding of pharmaceutical and veterinary products. AFNOR defines pesticides as being "substances or preparations that fight against enemies of cultural and harvested products". Pesticides include insecticides, fungicides, bactericides, herbicides, rodenticides, nematicides, molluscicides and other less important products. Almost 90% pesticides used for crop protection are made by human from non-toxic ingredients. The major groups of pesticides available are organochlorine, organophosphorus and organosulphur. They are further grouped into biodegradable and non-biodegradable depending on their life. The organophosphorus and organosulphur compounds do not accumulate in the environment as natural system like plants, bacteria, water and weather detoxifies them. The organochlorine compounds, especially chlorinated biphenyls and benzenes have a remarkable capacity to outlive the natural detoxification. DDT and its derivatives, BHC, endrin, aldrin, etc. The other toxic residues like phenolic compounds, hydrocarbon and surfactive agent (alkyl benzene sulphonates) are also released into the environment and they are biodegradable. They have a definite life span before being detoxified. Pesticides can be divided into several dozen classes by using general criteria for chemical affinity. Some important categories are:

3.2.3.2.1. *Organochlorinates*

Organochlorinated pesticides are characterized by their high chemical stability (absence of easily hydrolysable functional group). They are insoluble in water, but they are soluble in fats and organic solvents. They accumulate in animal fats. Their stability has led to an accumulation of residues, the progressive pollution of a number of natural media and a concentration in living organisms through intermediary food chain. They exhibit cumulative toxicity. Eg. HCH (lindane), DDT, HCB (hexachlorobenzene), Aldrine, Heptachlore, Chlordane.

3.2.3.2.2. *Organophosphates*

Organophosphate pesticides are generally more toxic (cholinesterase inhibitors) than other group, but they are more easily hydrolyzed and are thus degradable. This is the chief reason for the growth in their use. eg. Demeton S (-thiolophosphates), Parathion (-thinophosphates), Mevinphos (-phosphates), Analathion (-dithiophosphates)

3.2.3.2.3. *Polychlorobiphenyls (PCB)*

They contain two aromatic nuclei that bear a variable number of chlorine atoms. They may be diphenyls, terphenyls and higher polyphenyls or mixtures of these compounds. They are of high chemical and thermal stability; they are insoluble in water but soluble in lipids. They form complex mixtures of different compounds and a number of isomers.

3.2.3.2.4. *Other important pesticides*

The other important pesticides contaminated with aquatic food are Carbamates (Carbaryl), Dithiocarbanates (Sodium methyldothio-carbamate), Organometallics (Organomercurics, organostannics), Phenyoxyalkanoic acid (2, 4-D), and Triazines (Simazine).

Health hazards due to pesticides

Environmental chemical contaminants and pesticides from fish pose a potential human health hazard. Fish are harvested from waters that are exposed to industrial chemicals, pesticides and toxic elements. These contaminants may accumulate in fish at levels that can cause illness. The hazard most commonly associated with long-term exposures and illnesses associated with a single exposure are very rare. Concern for these contaminants primarily focuses on fish harvested from freshwater, estuaries and near-shore coastal waters rather than from the open ocean. Pesticides used near aquaculture operations may also contaminate fish. Intensive use of pesticides is an undisputed factor in food contamination and pollution of the food chain due to their residues. Toxic residues released into the atmosphere mix with rainwater before merging with rivers and finally seas. The final destination of these residues is the marine water and this is a cause of great concern. Heavy metal and chlorinated pesticide residues as a group of compounds are not degraded by natural mechanisms. Metallic elements are neither created nor destroyed and their total level in the universe remains the same, but due to human activity these deposits get redistributed and mix with marine water. The toxic residues enter into microscopic and macroscopic aquatic organisms through the food chain. Depending on the life span and feeding habits, the larger organisms accumulate the toxic residues at a higher and faster rate than other organisms due to their filter feeding habit. Hence, bivalves are selected as sentinel organism for early detection of water pollution. Analysis of these aquatic organisms showed that lower levels of these environmental residues accumulate in the body. In some instances, toxic levels of mercury and pesticide residues were also detected. The production and use of chlorinated pesticides is still on in India. Countries like USA and UK have realized this problem and banned the production and use of chlorinated pesticides. However, in India, the production of DDT and related compounds are in progress and these are widely used for malaria eradication programme.

Permitted pesticide residues

Federal tolerances, action levels, and guidance levels are established for some of the most toxic and persistent contaminants that are found in fish. In the case of molluscan shellfish, State and foreign Government Agencies, called Shellfish Control Authorities, consider the degree of chemical contamination as part of their classification of harvesting waters. The Guidelines of FDA, EU and India on pesticide residues are given in Tables 7, 8 and 9.

Table 7. FDA Guidelines on pesticide residues

Sl. No.	Deleterious Substance	Level	Commodity
1.	Aldrin/Dieldrin	0.3 ppm	Fish
2	Chlordane	0.3 ppm	Fish
3.	Chlordecone	0.3 ppm 0.4 ppm	Fish Crabmeat
4.	DDT, DDE, DDD	5.0 ppm	Fish
5.	Diquat	0.1 ppm	Fish
6.	Fluridone	0.5 ppm	Finfishcrayfish
7.	Glyphosate	0.25 ppm3.0 ppm	Finfish Shellfish
8.	Heptachlor/Heptachlor Epoxide	0.3 ppm	Fish
9.	Mirex	0.1 ppm	Fish
10.	Polychlorinated Biphenyls (PCB's)	2.0 ppm	Fish
11	Simazine	12 ppm	Finfish
12	2,4-D	1.0 ppm	Fish

Table 8. EU Guidelines on pesticide residues

Sl. No.	Pesticides	MRL (in ppm)
1.	DDT/DDE/DDD	5.0
2.	Aldrin & Dieldrin	0.3
3.	Polychlorinated biphenyl (PCB)	2.0
4.	BHC	0.3
5.	Chlordane	0.3
6.	Heptachlor/Heptachlor epoxide	0.3
7.	Mirex	0.1

Table 9. National Guidelines on pesticide residues

Sl. No.	Pesticide	MRL (in ppm)
1.	BHC	0.3
2.	Aldrin	0.3
3.	Dieldrin	0.3
4.	Endrin	0.3
5.	DDT	5.0

Control measures

To avoid health hazards, regular monitoring of the level of environmental residues in fish/ shellfish is the need of the hour. Many Nations and International organizations have drawn suitable criteria for acceptability of fish and fishery products based on the tolerance limits for each environmental toxic residue. Preventive measures for environmental chemical contaminants and pesticides include:

1. Making sure that incoming fish have not been harvested from waters that are closed to the commercial harvest of that species due to environmental chemical contaminants or pesticides

2. Receipt of the aquacultural grower's lot-by-lot certification of harvesting from uncontaminated waters, coupled with appropriate verification

3. Review, at time of receipt of aquacultured fish, of environmental chemical contaminant and pesticide test results of soil and water or fish flesh samples, and monitoring of present land use practices in the area immediately surrounding the production area

4. On-farm visits to the aquacultural grower to collect and analyze soil and water samples or fish samples for environmental chemical contaminants and pesticides, and to review present land use practices in the area immediately surrounding the production area

5. Environmental chemical contaminant and pesticide testing of fish flesh at time of receipt

6. Receipt of evidence that the producer operates under a third party-audited quality assurance programme for environmental chemical contaminants and pesticides

Methods for the determination of pesticide residues

1. General multiresidue method
2. Gas chromatographic method

3.2.3.3. Heavy metal contaminants

In addition to the above chemical hazards in aquatic foods, the heavy metal contaminants also play an important role in the safety of aquatic food. With the increase in industrialization, urbanization and other anthropogenic activities, pollution of water bodies is indeed a matter of concern. The human activities near coastal waters have caused deterioration of water bodies through trace/ heavy metal contamination. Though there is no precise definition for the term heavy metal, it can be defined by its density, which in this case is above $5g/cm^3$. Heavy metals have ecotoxicological importance due to their persistence and toxicity. One of the important characteristics of heavy metals is that they are non-biodegradable. In the aquatic environment, the heavy metals are redistributed throughout water column, deposited in sediments and consumed by biota. Excessive concentration of heavy metals in humans has been associated with the etiology of cardiovascular, renal, neurological and bone diseases. Regular monitoring of heavy metal concentration in the environment and food chain is important to prevent health hazards. Heavy metals that pose more threat to human health are mercury (Hg), cadmium (Cd), lead (Pb) and arsenic (As). The IARC classified as Hg, Cd and As carcinogens.

3.2.3.3.1. *Mercury (Hg)*

Mercury is distributed in the environment and is non-essential and toxic to the human body. Mercury exists in various forms such as elemental mercury, inorganic mercury and organic mercury compounds. Methylmercury (MeHg) and ethylmercury are common organic forms of mercury combined with carbon. Methylmercury is also formed from methylation of inorganic mercury by microorganisms in the environment and constitutes the major source of exposure from aquatic food. Methylmercury compounds are possibly carcinogenic to humans, Group 2B of IARC classification. Mercury in contaminated water has the potential to enter the food chain and once in the food chain, it bioaccumulates causing adverse effects to human health. Fish appears to be the primary source of MeHg poisoning in humans. MeHg levels tend to be higher in larger predatory fishes like tuna, king mackerel, swordfish; intermediate in medium-sized predatory fishes like snapper, trout; and lower in smaller fishes such as shrimp, clams or short-lived fishes like salmon. Once ingested, the gastrointestinal tract absorbs approximately 95% of MeHg. Because urinary excretion of MeHg is negligible, MeHg is primarily eliminated from the body in an inorganic form through the action of the biliary system at the rate of 1% of the body burden per day. Approximately 90% is excreted in stool and 20% of methyl mercury is excreted in breast milk from lactating women. Although the nervous system is the primary repository for mercury exposure, the systemic distribution has the potential to cause symptoms in different organ systems causing cellular, cardiovascular, haematological and pulmonary effects, effects on digestive and renal system, on the endocrine, reproductive system and also affects the foetus.

3.2.3.3.2. *Cadmium (Cd)*

Cadmium is one of the most toxic elements to which humans are exposed. It is classified as a human carcinogen as per IARC of WHO. Environmental exposure mainly occurs by contact with tobacco smoke, water and food. Shellfish is the highest contributor of Cd for non-smokers. It is widely distributed throughout the body, but the highest levels are found in liver and kidney. Recent studies suggest that Cd exposure may result in an altered DNA methylation pattern. Evidence suggests that exposure to cadmium does not induce direct DNA damage; however it induces an increase in reactive oxygen species (ROS) formation, which in turn induces DNA damage and can also interfere with cell signalling. More important seems to be cadmium interaction with DNA repair mechanisms. It is of particular importance that these effects are observed even at exposure to low Cd concentration that may be relevant to human exposure. Cadmium is able to pass to the foetus via the placenta affecting its central nervous system. Increasing evidence has demonstrated that Cd is a possible etiological factor of neurodegenerative diseases, such as Alzheimer's disease and Parkinson's disease. High levels of Cd in urine have been associated with higher overall mortality and increased risk of cancer. In addition to its nephrotoxic effect, Cd is also associated with increased risk of developing diabetes. Exposure to Cd also increases the risk of high blood pressure and it is considered a risk factor for cardiovascular mortality and morbidity.

3.2.3.3.3. *Lead (Pb)*

Lead is a soft and malleable metal, which is considered a heavy metal. It is a public health problem due to its adverse effects, mainly affecting the central nervous system in the most vulnerable populations, such as pregnant and lactating women and children. Once it enters the blood stream, 99% of Pb binds with the erythrocytes. During the pregnancy, the level of lead in maternal blood stream increases, it crosses the placenta which poses a threat to the health of the foetus. Deficiencies in calcium, iron, and zinc have been identified as risk factors for toxic effects being calcium deficiency one of the main causes for the severity of the effects. Not only during the pregnancy but also during lactation, Pb can pose a threat to infants. Breast milk from mothers with current exposure to lead or mothers exposed by the redistribution of bone lead has been identified as a source of exposure to the infant. The presence of Pb in the human body causes damage to the nervous system through several mechanisms. There is also a growing evidence of antisocial behaviour linked to early Pb exposure. Research also indicates that cumulative environmental Pb exposure is neurotoxic to adults acting as a risk factor for accelerated decline in cognition.

3.2.3.3.4. *Arsenic (As)*

Arsenic is a metalloid that is naturally present in low concentrations in the environment. It can be categorised as organic or inorganic, depending on the presence or absence of a carbon bond and may be found in one of three oxidation states, "3, +3 and +5", the trivalent form being the most toxic. The toxicity of the pentavalent form results from its conversion to the trivalent form. The inorganic forms of arsenic are generally more toxic than the organic forms and are responsible for most cases of arsenic poisoning in humans. Arsenic is a well-established human carcinogen as per IARC of WHO. The health consequences of arsenic exposure include respiratory, gastrointestinal, haematological, hepatic, renal, skin, neurological and immunological effects, as well as damaging effects on the central nervous system and cognitive development in children. The main routes of exposure to inorganic arsenic are ingestion of drinking water and inhalation of polluted air and dust, the former being the most important route in the case of millions of children in countries with high levels of arsenic in water. The organic compounds normally present in aquatic food are considered nontoxic or of low toxicity. Organic arsenic, in turn, is mainly found in fish and seafood in the form of arsenobetaine and arsenocholine. After ingestion, about 60–90% of organic and inorganic forms alike are absorbed into the blood stream from the gastrointestinal tract.

Arsenic needs to go under methylation as part of its biotransformation, which was previously considered detoxification, but is now known that the metabolite it produces is more toxic than the form exposed to previously. Monomethylarsonic acid MMA(V) and dimethylarsonic acid DMA(V) are the end products of arsenic metabolism. These are excreted through the urine. This balance between intake and excretion will determine the load of arsenic in the system. One of the principal mechanisms of arsenic toxicity is the induction of a strong oxidative stress with production of free radicals in cells that induce DNA damage, lipid peroxidation and decreased glutathione levels. The oxidative stress induced by chronic exposure to inorganic As is related to cytotoxic and genotoxic effects in the cells, acting as cause in the pathogenesis of diabetes, cardiovascular and nervous systems disorders. The inhibition of DNA repair processes is considered the main mechanism of genotoxicity. Cancer in skin, liver and kidney may originate from the molecular damage to proteins, lipids and DNA. Arsenic can be transferred from the mother to the foetus through the umbilical cord. That could create some adverse birth outcomes, increasing infection and risk for neurobehavioural impairment, cardiovascular disease and high blood pressure during childhood. These studies refer to children and mothers exposed to arsenic from water supplies. Nevertheless, even from seafood consumption children and infants may have higher exposure to metals because they consume more food in relation to their body weight and absorb metals more readily than adults.

Cephalopods (squid, cuttlefish and octopus) and crustaceans (shrimp and crab) are important class of aquatic food in commercial fishing. Cephalopods are preferred for their rich taste; low saturated fat and significant levels of minerals such as sodium, calcium, potassium, magnesium, iron, zinc, copper and phosphorus. Cephalopods have short life span, high growth rate and can accumulate high concentration of cadmium and lead. Shrimps and crabs have great importance for its taste, food value and nutrition. Crabs and shrimps being invertebrates tend to accumulate more metals when compared to fish due to specific differences in evolutionary phylum coping strategies. The accumulation of mercury is high in fast swimming fishes like tuna, saury, etc. The liver/ hepatopancreas is the main organ for accumulation in all the species. Cadmium is the metal of concern in both crustaceans and cephalopods, as the liver contains as high as thirty five times higher than that prescribed by Codex Alimentarius Commission.

3.3. PHYSICAL HAZARDS

The physical hazards include any potentially harmful extraneous matter not normally found in food. The extraneous matter found in aquatic food products can be divided into two categories. They are non-food safety hazards (e.g. filth), and food safety hazards (e.g. glass, metal, wood, bones, stones, hard plastic). The adverse health effect of physical hazards may be choking, injury including laceration and perforation of tissues in the mouth, throat, stomach or intestines. Broken teeth and damage to gums may also be the result. The FDA Health Hazard Board has found that foreign objects that are less than 7 mm maximum dimensions rarely cause trauma or serious injury except in special risk groups such as infants, elderly or surgery patients. Although physical hazards rarely cause serious injury, they are among the most commonly reported consumer complaints, because the injury occurs immediately or soon after eating, and the source of the hazard is often easy to identify. The control measures for physical hazards can include periodically checking all equipment for damage or missing parts passing the product through metal detection for metal inclusions, and passing the product through an X-ray detector for non-metallic objects.

CHAPTER 4

QUALITY PROBLEMS IN AQUATIC FOOD PRODUCTS

A quatic food products are generally grouped into eight categories viz. chilled, frozen, smoked, canned, bottled and similar products, salted, dried, marinades and other heat processed. However, few products are made into a single category and accordingly aquatic food products are divided into five major categories. They are chilled, frozen, canned, cured and other products. The control of quality is possible in each aquatic food product.

4.1. CHILLED FISH

It refers to unfrozen fish, which are brought from landing centre to sellers or processing factories and then distributed to retailers, using only chilling as a means of preservation. The processing applied is often only gutting, filleting, peeling or shucking. Large quantities of fish frozen at sea either in the whole form or as fillets are also thawed out on shore usually after intermediate frozen storage and thereafter is treated in exactly the same way as fish landed in the chilled form. The factors to be taken into account in controlling quality of the chilled products are basically the same as that of raw material. In particular, the pattern of spoilage is the same or very similar. It should be noted that the storage life of fish prepared from previously frozen and thawed fish raw material is, to all intents and purposes, the same as that prepared from equivalently fresh unfrozen fish raw material, always assuming that the freezing, frozen storage and thawing have been carried out properly. Fish in factory premises almost inevitably undergoes a good deal of further handling and is exposed to temperatures higher than ideal, and thus the potential for deterioration is greater. Fish cut into slices, steaks or fillets are more susceptible to spoilage because it has a greater surface area and thus, microbiological and other contamination are greater and the product tends to warm up more quickly. It is obvious that many faults in the quality of raw material cannot be corrected at the further processing stage. Fish subjected to unavoidable delays before or during processing can be reduced to or maintained at a low chill temperature most effectively by mixing it intimately with crushed ice. The use of bulk storage in chilled liquid media during either processing or distribution of products is often impracticable. It is normally impossible to reduce sufficiently rapidly the temperature of a solid mass of fish such as fillets by placing it in a refrigerated chill room. Such rooms should be used in conjunction with ice for the storage of chilled wet fish; their temperature should be at 2° to 3°C to allow a slow melting of the ice. Fish can sometimes be kept cool by immersing them in running water or being sprayed with cold water, where ice cannot be applied. However, prolonged immersion of fillets in water should be avoided. In addition to these measures, the product should not be subjected to high ambient temperatures or direct radiant heat.

Most fish is left unprocessed until after rigor mortis has passed off. But, sometimes fish may be filleted while they are in the pre-rigor state. Under these conditions, the process of rigor continues in the fillet, which leads to overall shrinkage, a roughening of the cut surface and a loss of fluid. In extreme cases, the length of fillet may shrink by 30 - 40%; lose of 25% of its weight and the surface becomes corrugated. This problem can be overcome by keeping them a day or so in the whole state until they are in or through rigor. Alternatively, if the fillets are to be frozen, the fish should be kept chilled throughout and handled expeditiously in order to prevent the fillets going into rigor. The ratio of ice to fish packed in containers for distribution should be adequate and most probably 1:1 to reduce the temperature of the fish quickly and to maintain it near 0°C throughout its transit. This ratio has to be adjusted to take into account the prevailing ambient temperature and type of container used. Ice should be intimately mixed in layers with the fish and packed in such a way that heat entering the container is absorbed before it reaches the fish. Wet strength paper is sometimes used to protect the fish from indentations caused by ice lumps and possible contamination. This practice is unexceptionable as long as it does not interfere with the cooling by the ice. The skilful use of crushed or flaked ice coupled with some kind of transparent protective cover that shield the fish from dirt are best for the retail display. Mechanically refrigerated chill cabinets or display slabs are an alternative, though they tend to dry the fish, because the ambient humidity is low and the control of temperature in them is also not always easy and sometimes damaging partial freezing of the product can occur. Cleanliness, hygiene and sanitation of premises and equipment should receive special attention. Guts, filleting and trimming offal, shell and other debris should not be allowed to contaminate the finished product and should be rapidly removed from its vicinity. It is often advisable to chlorinate the water supply in factories or processing areas.

It is possible to extend the storage life of fillets and peeled or shucked shellfish by chemical treatment, irradiation or packing in gas-tight plastic pouches either evacuated or containing CO_2. Chemical treatments include dipping in or spraying with solutions of antibiotics such as OTC or chelating agents such as ethylenediamine tetra-acetic acid (EDTA). There is also some minor use of sulphites to prevent blackening in chilled crustaceans. In practice, the disadvantages of these adjuncts or alternatives to chilling currently outweigh the advantages; and hence, few are used. Chilled fish especially in the form of fillets loses fluid or drip during distribution, storage and display. The phenomenon is most marked with fillets or steaks cut from fish previously frozen and thawed, resulting in a weight loss of up to about 5%. This kind of loss can be taken care of by adding sufficient initial overweight. The cut surfaces of fillets prepared from frozen thawed fish tend to have a poorer bloom and appearance than corresponding fillets cut from unfrozen fish. Moreover, the unsightly fluid can accumulate in the packages resulting in reduced customer appeal. All these quality losses can be corrected by dipping the fillets immediately after cutting in a 5-10% aqueous solution of polyphosphate (as sodium or potassium tripolyphosphates or sodium or potassium pyrophosphates or sodium hexametaphosphate) for 1-2 min. These substances act by swelling the outer surfaces of the fillet so partially sealing them. An additional benefit from this treatment is the glossy translucency imparted to the cut surfaces. A similar but poorer effect can be achieved by dipping in concentrated brine. But, this is not recommended because the fish often become too salty. Overtreatment with polyphosphate results in a slimy unpleasant product. There is a practice of treating fish and shellfish with flavouring agents, the aim being to restore something like the fresh flavour that has been lost as a result of handling, washing, poor freezing, etc. Citric acid, ascorbic acid and monosodium glutamate are added to small peeled crustaceans and hydrolysed vegetable-protein mixtures to white fish.

4.2. FROZEN FISH

Freezing is a type of partial, gentle dehydration in which the water is removed as ice. It is a means of arresting either partially or completely the deteriorative actions of microorganisms and enzymes. Microorganisms cease to multiply below about - 10°C and the activity of enzymes is in general rapidly reduced, as the temperature is reduced below the freezing point of about - 1°C. Freezing cannot reverse deteriorations that have already occurred. Fish with a certain degree of pre-freezing spoilage retains it throughout freezing, frozen storage and thawing. Controlling the quality of frozen fish includes selecting the quality raw material and controlling the quality of fish during freezing process. Whole round white fish stored nearer 0°C should be frozen within 2 - 3 days of death, if good quality thawed products suitable for subsequent filleting and processing are to be obtained. This period can be 5 - 6 days for most flat fish and only 1 - 2 days for small pelagic fish. Most shellfish in shell or not, after catching, should not be kept in ice for more than 2 to 3 days before freezing to obtain good quality.

Off-odours and off-flavours gradually develop during frozen storage of fish. The exact nature of these deteriorations depends on the species but most particularly on whether it is a fatty fish or not. The very characteristic off-odours and off-flavours assumed by lean fish and shellfish are variously described as acid, bitter, turnipy, cardboardy or musty, those of fatty fish are typically rancid, oxidized, painty or linseed oil-like. The fat on the surface of whole, thawed fatty fish when in an advanced state of rancidity feels gummy to the touch. The texture also gradually changes from the usual soft, moist, succulence of fresh or recently frozen fish to unacceptably firm, hard, fibrous, woody, spongy or dry. Copious amounts of exude can be pressed out after frozen storage under poor conditions. When fish that has deteriorated in cold storage is smoked, the attractive glossy pellicle formed during normal curing becomes increasingly difficult to obtain and eventually only a dull, matt product results. The necessary tendency to become sticky when ground with salt and to form a gel satisfactorily on steaming as in the manufacture of fish balls, fish sausage or kamaboko is also gradually lost and the fish becomes completely unfit for this purpose. All these textural changes are caused by an essentially irreversible phenomenon known as "denaturation" suffered by the flesh proteins. Appearance also suffers. White fish and shellfish become opaque and yellowish; fatty fish develop a 'rusty' appearance; the pigments of fish and shellfish tend to fade and become duller or change in hue; 'bloom' gradually disappears. The effects are enhanced, if the fish are allowed to dry out, which tends to happen naturally during frozen storage. Dehydration is a kind of quality loss, because product weight is lost but equally the surface and thin parts of the fish become irrecoverably dry and porous. This condition is known as "freezer burn", which renders products quite inedible.

The storage life depends upon initial freshness: the staler the fish before freezing the shorter is its shelf life during frozen storage. Because the effect of temperature is so pronounced, it is of great importance in maintaining quality to keep fish close to the lowest temperature reached after freezing. Wide fluctuations in temperature during storage should also be avoided, as these can lead to rapid migration of moisture with resulting damaging dehydration and enhanced deterioration. The signs of deterioration in still-frozen fish are sometimes not particularly noticeable and are totally hidden, if opaque packaging is used. Controlling of dehydration can be achieved by reducing or preventing the loss of moisture through either coating the frozen fish or product in a layer of ice or placing it in a more or less tightly fitting wrapper. The layer of ice is deposited by simply dipping the frozen product in or spraying or brushing it with clean water. This process, known as 'glazing' is a cheap and effective remedy. Under drying conditions in a cold store, the glaze evaporates rather than the moisture in the product. A glaze thickness of 0.5 to 2 mm, representing for most products a weight of 5 to 15% of applied water, is generally sufficient. If the glaze evaporates completely from any part of the product before the end of the storage period, it should be renewed. Products frozen as separate pieces like individually quick frozen (IQF) can be glazed as such, but blocks of whole fish or fillets are normally glazed on the outside after freezing is completed. The additional advantage by freezing fish and water together in a plastic or waterproof bag is the protection against physical damage to these vulnerable species. The efficacy of plastic bag depends on its degree of imperviousness to water vapour at low temperature and the degree to which the product is closely and completely enclosed. However, even under the best conditions, some slight moisture loss over long periods has to be accepted, because water tends to evaporate from the product surface and condense on the interior of the package to form a snow called as "in-package desiccation". Coatings such as batter also afford a useful measure of protection against dehydration.

Oxidative rancidity can be controlled by preventing oxygen from attacking the unsaturated lipids in the fish. The glazes used to protect against dehydration are also useful as a physical barrier between the product and oxygen in the air in protecting against oxidation and a doubling of storage life of a fatty species is typically obtained by thorough glazing. Packaging in oxygen-impermeable flexible films also offers practical extension of storage life, but, the oxygen in the package must be removed before storage either by evacuating the space between film and product or by replacing the air with an inert gas like nitrogen. Vacuum packaging is found to be more practical, though it is only applied to expensive small items like shrimps. Certain chemicals known as antioxidants have the property of inhibiting oxidative rancidity but they have to be mixed intimately with the lipid in the fish. Sodium chloride has a pro-oxidant action and as far as possible should not be mixed with or applied to fish destined for frozen storage. The textural deterioration caused by protein denaturation can be removed by means of controlling the temperature. Application of polyphosphates can also offer by a kind of sealing mechanism, some measure of protection against fluid loss or thawing. Similarly, problems of frozen storage of crustaceans can be avoided by cooking the animals before freezing. If frozen, raw lobsters tend to deteriorate rapidly and are difficult to shell after thawing. Cooking shrimps before freezing facilitates peeling post-thawing. Since fish of low pH tend to deteriorate faster than those of high pH, selection before freezing and frozen storage offers one means of controlling the situation. It has been found that mince containing blood or particularly tissue from kidney or viscera or liver deteriorates in flavour and texture very rapidly when frozen as blocks. In the preparation of frozen laminated blocks of fillets, the requirements of minimum incidence of bone, skin, scales, connective tissue, belly membrane, etc. should be observed. Tainting from pouch materials is possible in the case of boil-in the bag products; only tested and recommended films should be used. Frozen fish can be contaminated with refrigerants such as ammonia or trichloroethylene. At normal temperatures and on cooking, most of the refrigerant absorbed inevitably by the product evaporates, but traces always remain.

Common quality defects in frozen fish are black spot formation in shell-on shrimps, drip loss or weight loss during thawing, loosening of head in whole shrimps, dehydration or desiccation, blackening and browning in frozen lobster tails, yellow discolouration in lobster meat, squid tubes, cuttlefish fillets and pomfrets, bacterial problems, presence of foreign materials, presence of excessive numbers of inedible materials, green discolouration in frozen swordfish and ship-frozen tuna fillets and protein denaturation.

1. Black spot in shell-on shrimps

Black spot formation or 'melanosis' in shell-on shrimps is a major problem in the freezing industry. It is due to the action of tyrosinase enzyme on the tyrosine in the presence of oxygen and heavy metals, resulting in the formation of melanin pigment. It requires oxygen in addition to the presence of heavy metals like copper and iron. Melanosis does not interfere with the eating quality, but affects the appearance. Prevention of melanosis demands cutting of access to oxygen. In the usual practice, keeping the shrimps in ice with a layer of the same above the material prevents black spot formation. The enzyme is more concentrated in the head portion. Hence, removal of the head portion immediately after catch and washing of the tails followed by icing delays the phenomenon. A dip in 0.2-0.5% of sodium or potassium metabisulphite for 1-2 min. before packing is found to be effective in controlling the black spot formation. However, higher levels of sulfite cause bleaching of shell colour. Since there is a limit of 100 ppm for residue of sulphite as SO_2, it is presently recommended to use 4-heptyl resorcinol to prevent black spot formation.

2. Weight loss on thawing and thaw drip

On thawing, frozen seafood loss some weight, as thaw drip. Thaw drip usually consists of water containing soluble nutrients and flavour-bearing components. Excessive amount of drip reflects degradation in the quality of frozen products. If the drip is discarded, water-soluble nutrients and flavour are lost. Food that exudes a large amount of drip is usually dry and woody or tough in texture. A large amount of drip adversely affects the appearance of the product and causes a shrinkage or loss of weight. Weight loss due to thaw drip is to be compensated by addition of excess weight at the time of freezing. The extent of weight loss due to thaw drip depends on the type of seafood. The loss from frozen HL shrimps is 5%, frozen PUD or PD shrimps is 10-15% and cooked frozen shrimps is 7-10%. Thaw drip losses increase with pre-freezing ice storage period. Thaw drip losses are prevented by treatment with phosphates. However, the present food laws of the country do not allow use of phosphates. Hence, control of weight losses requires careful manipulations during draining, weighing, freezing and frozen storage.

3. Loosening of head in whole frozen shrimps

Careful handling of whole shrimps from the stage of catching is an essential step to reduce this defect. Fresh shrimps should be kept in very finely crushed ice and water nearer to 0°C from the time of catching and should be processed without much delay. Chlorinated water at 10 ppm level should be used for washing. A dip in 0.2% solution of sodium metabisulphite for 2 min. controls the loosening of the head.

4. Dehydration in frozen seafood

The white patches or 'freezer burn' on frozen fish meat is due to severe dehydration/desiccation. Proper packaging with exclusion of air pockets from the package, sufficient glazing, maintenance of constant storage temperature and relative humidity are essential for controlling this defect. The critical limit of dehydration (rate of weight loss) or water vapour transmission rate (WVTR) in frozen fish during cold storage is 50 $g/m^2/day$.

5. Blackening and browning in frozen lobster tails

Black spot development also occurs in lobster tails due to the action of enzyme tyrosinase or polyphenoloxidase. It is also noted at the cut ends and along the lining of the meat. Keeping the lobsters live up to the time of processing helps in controlling the black spot formation. Brown discolouration develops during frozen storage at the cut end of meat. Proper glazing, wrapping and low temperature storages are remedial measures.

6. Yellow discolouration in frozen lobster meat

Yellow discolouration in frozen lobster meat is due to oxidation of the red astacene to a yellow pigment. This discolouration is accompanied by the production of off-odours and off-flavours. This can be minimized by the application of antioxidants, maintaining at low temperatures and packing in vacuum condition.

7. Discolouration in squid tubes and cuttlefish fillets

Yellow discolouration of the squid tubes and cuttlefish fillets are the serious problem. Removal of the appendages, ink sac and gut contents, followed by washing and bleeding immediately after catch prevents yellowing. This dressed material has to be stored in ice and water until it is used for freezing.

8. Yellow discolouration and dehydration in frozen pomfrets

Yellow discolouration normally occurs in frozen pomfrets due to oxidation. There is also considerable loss in weight of frozen pomfrets, if stored unglazed. Dipping of frozen pomfrets in glazed water containing the antioxidant ascorbic acid (1%) or a mixture of ascorbic acid (2%) and citric acid (0.2%) for 2 min. has been found to prevent the yellow discolouration and dehydration.

9. Bacterial problems in frozen seafood

Frozen seafoods should be free from excessive bacterial load. The bacterial load in raw frozen seafood should not exceed 10^5/g. Some types of bacteria like *E. coli,* coagulase-positive *Staphylococcus aureus* shall be in a very limited number. *E. coli* in raw frozen seafood should not be more than 20/g; it should be absent in cooked frozen products. Coagulase-positive *Staphylococcus aureus* should not be more than 100/g. *Salmonella, Vibrio cholerae* and *Listeria monocytogenes* shall be absent in the frozen products. Food standards are generally strict about bacterial quality of precooked products. Precautions and sanitary practices like chlorination of water supplies both for use in processing and ice manufacture, application of regular cleaning schedules, workers' hygiene, etc. are very important to keep down bacterial contamination.

10. Presence of foreign materials

Foreign materials such as flies, fibre pieces, hairs, bits of paper and excessive sand should not be present in the frozen products. More care at the time of packing and exclusion of flies in the processing is very much essential. Avoidance of peeling of shrimps on floor, use of pond water for washing, stoppage of hut-peeling and proper washing of whole shrimps can help in reducing the sand content.

11. Presence of excessive numbers of inedible materials

The presence of materials like veins, shell pieces, antennae, etc. in shrimps is objectionable and causes inconvenience to the consumer. Proper sorting and washing are the remedial measures.

12. Green discolouration in frozen swordfish and tuna

Green discolouration is a common defect in frozen fillets of tuna and swordfish. Hydrogen sulphide derived from the deterioration of fish meat reacts with hemoproteins in the presence of oxygen to form sulfhemoproteins such as sulfhemoglobin and sulfmyoglobin. These products are green pigments, which exhibit an absorption maximum at 618 nm and cause the discolouration. A sour smell accompanies the occurrence of green colour. The off-flavour associated with greening is found to be due to H_2S and isovaleric acid. The production of H_2S in meat is a result of the growth of bacteria like *Pseudomonas mephitica, Alteromonas putrefaciens* and *Proteus vulgaris.* Discolouration of the meat could be prevented by icing the fish and keeping them cold until they could be filleted and frozen or by means of proper bleeding.

13. Protein denaturation

During frozen storage, fish flesh proteins undergo denaturation due to the fluctuation in the temperature of cold storage. The original structure of the protein is destroyed resulting in the loss of functional properties such as water holding capacity, gel forming ability, etc. It also affects the texture of the fish. It can be prevented by controlling the temperature of the storage.

4.3. CANNED FISH

Canning is the heat processing of food in a hermetically sealed container in order to reduce the effect of bacterial contamination to a commercially safe level. The basis of canning process rests with the destruction of the organisms, which are already present in the raw material, by heat and prevention of the entrance of others. The organoleptic and nutritive properties of the product are also retained to the greatest possible extent by canning. Its preservative action depends upon the heat inactivation of intrinsic and microbiological enzymes and the protection of the product from subsequent attack by microorganisms and atmospheric oxygen. The hermetically sealed container used also protects against damage and contamination with dirt. Control of quality again depends primarily on selection of good quality raw material. The canning process may sometimes give highly acceptable product from raw material that possesses slight, incipient spoilage features. But, it is better to avoid raw material of very high bacterial load. When the fish is too fresh, i.e. in rigor, it is often difficult or impossible to handle in the stages preparatory to canning; very fresh tuna is more liable, than less fresh, to slow 'scorching' or darkening of the meat surface exposed to the headspace. Fatty, pelagic species are by for the most commonly canned and their fat content often determines the quality of the end product. The fat content of the raw material should be between 7 and 15% to make the best quality canned sardines and sprats. The quality defects, which are dependent upon compositional differences in canned fish, are 'greening', colour variations and 'struvite' formation. The greening is a grey or greenish-grey colour in tuna that detracts from the normal buff or pinkish tan colour. The cause is not entirely clear, but is connected with changes in the haem pigments. Thorough bleeding and hence reduction in the concentration of haem pigments after catching tends to reduce the incidence. The colour is a major compositional factor influencing consumer acceptance in canned tuna and salmon. The shade must be characteristics of the species traditionally canned and as uniform as possible with and between batches of finished product. This is achieved by visual inspection of raw material by trained personnel. 'Struvite' is hard, glassy crystal of magnesium ammonium phosphate formed in the period after heat processing from natural constituents of the fish. The incidence of struvite in the tuna, especially albacore (*Thunnus alalunga*), is associated with high pH. Measurement of pH on the raw material and selection for canning in the solid form of fish having pH values below 6.0 eliminates the defect. Struvite in other canned species can be reduced by either lowering the pH of the contents with citric acid before canning or adding about 0.5% sodium hexametaphosphate or sodium acid pyrophosphate.

The primary objective of using correct retorting times, temperatures and procedures is to ensure that the most heat resistant microorganism (*Clostridium botulinum*) in the pack killed. In practice, it is not possible to achieve 100% or absolute sterility in all containers without excessively damaging the product and the aim is to reduce the risk of having pathogens present to an extremely low level while producing a food that keep for a long time at ordinary temperature without spoiling. Cooling after heat processing should be carried out rapidly to avoid over processing and to reduce the risk of struvite formation. A form of over-processing ('stack burn') occurs, if cans are packed while still hot into outer cartons or closely together in piles. Under these conditions, cooling can be very slow. The remedy is to cool thoroughly before stacking or packing. Pressure cooling may be advantageous. Potable quality water should be used for cooling and washing. Traces of water are prone to be sucked into the can during these stages and if microorganisms are so introduced, spoilage may ensure; chlorination of plant water helps to eliminate this possibility by greatly reducing the numbers of microorganisms. The seam in containers can be weakened and broken by rough treatment and contaminating microorganisms gain access to the contents. Thus, care in handling filled containers is necessary at all times. It is good practice to emboss cans or mark containers with a code showing a few essential details of production so that defective batches may be readily identified, isolated and if necessary, recalled. Coding also makes it possible in principle to trace back the cause of defectiveness.

Microbiological deteriorations arise from insufficient heat processing during retorting or from contamination through seam at any stage up to detection. The result is almost invariably a foul smelling product and often a swelling or bursting of the can due to microbiological generation of gas. Rarely in the case of fish products, microbiological spoilage may not result in the generation of gas and development of pressure. Various kind of deformity with special names such as flipper, springer, soft swell and hard swell is caused by varying degrees of spoilage. Chemical spoilage arises typically from the gradual attack of substances in the product on the metal of the can. Commonly these substances are the acids in sauces or packing medium, the result being internal corrosion, generation of hydrogen with swelling of the can and discolouration of the contents. In addition, traces of metal are leached from the tinplate, contaminating the product and so possibly causing a health hazard. This form of deterioration is virtually eliminated by coating the interior of the can with a protective lacquer such as zinc oxide or by using aluminium cans. A second form of chemical spoilage, which does not affect the fish itself, is internal blackening of tinplate cans. Volatile sulphur compounds derived principally from the crustaceans can gradually react under some circumstances with metals in the plate to form a black or dark disfiguring deposit of iron sulphide or tin sulphide. Much of the problem can be overcome by using tinplate coated with special lacquers, such as zinc oxide, which still reacts with sulphides, but it gives a white unnoticeable compound, zinc sulphide. Further protection is afforded by interposing parchment paper between the fish and the can interior. Aluminium cans do not suffer from this problem. A similar phenomenon sporadically affects the packed meat itself, particularly with canned shrimp and crab. The sulphur compounds react with iron and probably also copper present in the meat. The metals can be either of natural occurrence or picked up during preparation. The addition of small amounts of citric acid, citrate or phosphoric acid to the final filling medium or dipping the shellfish in solution of these substances before canning inhibit the release of sulphur compounds and act as chelating agents for the metals. Both microbiological and chemical spoilages often take some time to manifest themselves and hence it is of advantage to store cans for a few weeks or months before final inspection and despatch.

The correct weight of fill and drained weight of fish must be complied with the requirements, taking into account of allowable tolerances. Similarly, the species, size or count of the contents should meet the requirements. A defect that occurs sporadically and unexpectedly in skipjack tuna (*Katsuwonus pelamis*) and albacore is an orange discolouration. Like 'greening', this is not detectable in the raw material and the cause is unknown. End product inspection is the only way of detecting its occurrence. When badly frozen and thawed mackerel and salmon are used as raw material for canning, the finished product is sometimes rendered unattractive in appearance by the formation of clumps or layers of coagulated protein called as 'curd'. It is prevented by dipping the fish before canning in brine or a solution of tartaric acid, which seals the cut surfaces and prevents the escape of fluid containing potentially coagulable protein. A grey or bluish discolouration arising from blood pigments and possibly the presence of iron occurs principally in the relatively small amount of canned crab. The defect can be atleast partially overcome either by cooking at a moderate temperature which coagulates the meat protein, but allows the still fluid blood pigments to drain away or by treating the meat before cooking with brine containing a small amount of aluminium sulphate or EDTA. Corrosion may affect the exterior of tinplate cans or metal fastenings. Condensation of moisture or inadequate washing are the main causes and should be guarded against, by providing adequate ventilation and an evenly controlled temperature of storage. Honey combing formation in canned tuna is another defect. This is due to the use of stale meat and physically damaged meat which results in the muscle tissue gives out its water and when cooked it evaporates and so hollow spaces are seen in the meat, which is similar to honey comb. Greening occurs as a result of oxidation reaction between TMAO and meat pigments such as myoglobin during heating in the presence of cysteine. It may give a bitter taste. High TMAO content is found in the tail end of tuna meat. Stale fish contains more TMAO. TMAO is responsible for greening in canned tuna. If the TMAO level in the raw meat is less than 8 mg%, there is no greening in the canned product. If it is above 13 mg%, greening occurs in the product. If it is between 8 and 13mg%, greening may or may not occur in the canned product.

The chief causes of microbial spoilage in canned products are underprocessing, inadequate cooling, leakage through seams and pre-process spoilage.

i. Underprocessing

Any pack, which suffers spoilage as a result of the activities of microorganisms surviving the heating process, can be termed as underprocessed. In underprocessed packs, the activity of the surviving microorganisms may result in gas production, which causes the can to become a 'swell' (which is due to the growth of *Clostridium botulinum*) or the contents may undergo acidification or some other undesirable changes affecting quality, but gas is not produced. When growth of microorganisms occurs without gas production, the affected cans have a normal appearance externally and spoilage is only detectable after can has been opened. Such spoilage is called as flat sour spoilage, which is due to the growth of *Bacillus stearothermophilus*.

ii. Inadequate cooling

Some bacteria such as thermophiles multiply rapidly in a high range of temperature and failure to cool cans immediately after processing to a temperature of about 35°C may lead to serious spoilage.

iii. Leakage through seams

The microorganisms contaminating canned fish as the result of post-process leakage of the containers may be of widely varying types. The main source of the organisms is the cooling water. In ideal seaming, the edges of the cover and the body are properly folded to form five folds of metal. In improper seaming, the cover hook is compressed against the folded body hook creating an opening between the body and the cover for a portion of the perimeter through which spoilage microorganisms enter the can.

iv. Pre-process spoilage

Spoilage of this type is an example of faulty canning practice, whereby bacterial development in the fish is permitted during the preparation period. Processing may subsequently sterilize the pack, but the liberation of gas produced by the organisms during the lag period before processing may cause swelling or flipping of the cans.

Common spoilage in canned products are blue / black discolouration in canned shrimp and crab meat, underweight in canned shrimps, bacterial problems in canned products, excessive water content in oil packs and high sand content in canned mussels and clams.

1. Blue / black discolouration in canned shrimp and crab meat

Shrimps canned in brine are known to develop blue / black discolouration at the head portion of the meat. This is mainly due to formation of copper and / or iron sulphides. Copper and iron are present in very small quantities in the shrimp tissue and H_2S is formed from the sulphur content in the amino acids of protein during processing. Discolouration occurs when the concentration of copper and / or iron exceeds certain critical levels. The muscle takes up copper and iron during processing from contact surfaces, water, ice, salt, citric acid, etc. and they combine with sulphur present in shrimp to form sulphide. Discolouration due to copper and / or iron sulphides can be prevented by adjusting the titrable acidity in the fill brine and also by addition of EDTA disodium salt in the fill brine at the rate of 50 mg%. Very fresh shrimps require only lower quantities of citric acid to adjust the titrable acidity to 0.06%, whereas raw material stored in ice requires more acid. Crabs too contain higher levels of copper. Blueing can be prevented by either proper bleeding at the time of butchering or by addition of EDTA salt.

2. Underweight in canned shrimps

Canned shrimps should have moisture content of 72%, which is called as equilibrium moisture level. To get correct drained weight, shrimps should be blanched to the equilibrium moisture content. If the moisture level is higher and the material is underblanched, more water is removed from the meat during processing leading to the underweight of the canned product. Hence, it is always essential to standardize blanching conditions and to follow it strictly. Particular attention is required to keep up the concentration of salt and citric acid in the blanching brine, as also the time of blanching to obtain correct drained weights.

3. Bacterial problems in canned products

All canned products are to be commercially sterile. Commercial sterility means that though some bacteria may be present in the can, they may not be able to multiply and spoil the contents of the can under the environmental conditions available in the can. Sterility (commercial or absolute) is tested after incubation of the cans at 37°C or 56°C for 14 or 4 days, and examining for growth in nutrient media. As a whole, 3% of canned shrimps are shown to be not commercially sterile. Bacterial problem may be due to insufficient sterilization and postprocess contamination from cooling water. It is essential to see that the processing time and temperature are kept up correctly. The processing line needs to be kept in clean condition and there should not be any hold up. Postprocess contamination usually occurs from cooling water. Cooling water shall be of sound in bacterial quality and shall contain free available chlorine. Contamination can take place through faulty seams as also through normal seams. Seams, though normal, can always allow sucking in of water at the time of cooling because of the strain due to temperature differences. The preventive step for faulty seam is checking of seam after every 200 cans.

4. Excessive water content in oil packs

Fish canned in oil shall not contain more than 10% water in the drained liquor. More water in the fill medium is a defective quality factor as it amounts to cheating the consumer and lowering the shelf life considerably by corroding the can and causing deterioration of the contents. Proper precooking and draining of the cook drip prevent the excess water content in the pack.

5. High sand content in canned mussels and clams

It is a serious defect in canned clams and mussels. Removal of sand from the raw meat during processing is particularly difficult. The live material shall be subjected to purification in bacteriologically clean water for 24 - 48 h, a process called depuration and so that there is an appreciable reduction in the bacterial load and sand content of the meat.

4.4. CURED FISH

Smoking is a method of preservation effected by combination of drying and deposition of smoke constituents. When fish is smoked, it is subjected to basic treatments viz. brining, drying, smoking and heat treatment. Formaldehyde, carbonyls, furans, esters, lactones, alcohols, phenols particularly polycyclic aromatic hydrocarbons (PAH) and acids are the important constituents of smoke involved in smoke curing of fish. Among these, phenolic constituents are supposed to be most effective in preserving fish. Benzopyrene, a PAH, is highly carcinogenic. Formaldehyde is an anti-fungal agent. Phenol has antimicrobial and anti-oxidative effect. During hot smoking, the temperature of smoke may rise; at times to above 100°C while the flesh reaches 60°C and gets cooked. The heat treatment also results in partial sterilization, although subsequent re-infection and spoilage of the cooked flesh is still quite rapid. During cold smoking, the temperature is not more than about 30°C and the fish is not even partially cooked. The process of smoking and the brining that usually preceeds it can not entirely mask spoilage or defects that are present in the raw material. Therefore, the quality of smoked fish depends to a large degree on that of the raw material. The rise in temperature during smoking may volatilize some of the objectionable compounds present in spoiling fish. The characteristic yellow, golden brown or brown colours imparted to the fish by smoke constituents or artificial smoke colours are capable of disguising minor imperfections of colour in the fish. Sometimes yellow or brown dyes are used before smoking to impart an attractive deeper colour than is obtainable by normal smoking. If spoiled fish is used, there is a danger that the food dye may lose its colour or even turn pink with obvious disastrous effects on quality. The change in colour is caused by the high pH of the spoiled fish. The strong flavours introduced by salt and smoke constituents are capable of concealing some incipient spoilage flavours. Apart from freshness, the quality of smoked pelagic fish depends to a considerable degree on the initial fat content. Fish meat having low fat of 1% and below are thinner and have a less glossy, attractive appearance and also tend to taste drier and less succulent than those of high fat content.

Though the spoilage of smoked fish at chill temperature is basically similar to that of unsmoked fish, the intrinsic and microbiological enzyme changes are modified to some extent by the presence of salt and smoke. The partial or complete cooking given in the hot smoking process also modifies the spoilage pattern, though the end character is still ammoniacal, faecal and sulphide. Low moisture content of smoked fish favours the growth of moulds (*Aspergillus* spp.) that originate in the wood used for smoking; the product, especially after wrapping, may therefore suffer additional musty flavoured deterioration from these microorganisms. On storage, the fresh aromatic smoke flavour of newly smoked fish becomes weaker, blander or unpleasantly tarry. Rancidity is much more noticeable in stale smoked fatty fish than in unsmoked fish. The combined effects of salt, smoke constituents and the drying that accompanies the smoking process are significantly preservative and the storage life of smoked fish is always longer than that of the unsmoked fish held under the same conditions. Heavily brined, smoked and dried fish remain edible for several weeks at 0°C and for several days at 15° - 20°C. The modern tendency is to reduce reliance on heavy curing as a means of preservation and to chill or freeze the lightly smoke flavoured product. The main factor in controlling deterioration in smoked fish during storage is temperature. Fish after smoking is warm and should be cooled in cold air as rapidly as possible to avoid unnecessary spoilage. Packing warm products closely can lead to 'sweating', that is the tendency of moisture to come to the surface and render it sticky. The production of good and consistent quality smoked fish basically depends upon maintaining certain levels of moisture (10-15%), fat, salt (15-20%) and occasionally dye content and of degree of smoke deposition. A good quality smoked product should possess a smooth glossy pellicle; a dull, ragged or gaping surface severely detracts from appearance. The production of a good gloss depends upon correct brining, adequate draining after brining and correct drying in the smoking kiln. Stale, poorly cold stored, poorly cleaned and washed or pre-rigor fish give poor gloss. In order to obtain a satisfactory smoke flavor, the correct uncontaminated wood or other cellulosic material must be used; some woods impart unpleasant resinous or acrid flavours. Checks should be made to ensure that the wood blocks, chips or sawdust have not been treated with preservatives and do not contain glues or remnants of plastics. It is possible to avoid the use of the smoking process altogether by adding smoke-flavouring mixtures such as liquid smoke to the fish.

Browning and tanning are the quality defects in smoked fish. Browning of meats is caused by the thermal decomposition products arising from macromolecular carbohydrates during smoke generation. During surface treatment of fish with smoke, formation of a secondary skin is observed. It is called as tanning effect. The reactions of carbonyls and proteins are mainly responsible for the tanning effect. Microorganisms responsible for spoilage in smoked fish are moulds especially xerophilic moulds, which can even grow at a water activity (a_w) level of 0.65. *Aspergillus glaucus* group including *A. amstelodami, A. chevalieri* and *A. ruber, Aspergillus flavus, A. restrictus, A. candidus* and *Penicillium* sp. normally encountered in smoked fish. Commonly used preservatives in smoked fish products are acetic acid, benzoic acid, propionic acid and sorbic acid. Propionic acid is particularly effective against moulds, while sorbic acid is more generally effective and is used at levels of 0.02 - 1.6% in various foods. Some moulds capable of growing on smoked fish may also produce mycotoxins. Certain toxigenic strains of *A. ochraceus* can produce the toxins penicillic acid and ochratoxinA in dried foods. *A. flavus* and the closely related *A. parasiticus* can produce a toxin known as aflatoxin even in cured fish and shrimps. Dangerously high levels even up to 600 - 700 ppm of aflatoxin (B1 and G1) has been found in cured fish.

Salting is a method of preservation based on the penetration of salt into the tissues and governed by the various physical and chemical factors such as diffusion, osmosis and a series of chemical and biochemical processes associated with changes in various constituents of fish. Salting starts the moment fish surface comes into contact with salt. The end of the salting process is the moment when the entire fish has reached the required salinity and acquired the appropriate taste, consisting and odour. Three kinds of salting are dry salting, wet salting and mixed salting. Success in making a good quality product invariably depends upon achieving in the early stages a rapid enough increase in salt concentration and in accompanying dehydration to prevent spoilage. Salting is a fairly slow process often conducted at ambient temperature. Salt penetration occurs from the outer layers so that for some period the inside of the fish remains unsalted and of normal water content. If neither salt penetration nor rate of dehydration is sufficiently rapid, microorganisms on the inside multiply and spoil the fish. To prevent this, the salt needs to be brought into as intimate contact as possible with the fish flesh like splitting large fish to give an enlarged surface area from which the salt can diffuse to the centre of the flesh, and pressing in stacks to aid rapid removal of water. All types of white fish can be dry salt cured, though elasmobranchs may give trouble with evolution of ammonia or amines unless the salting and drying processes are carried out quickly. Pelagic fish of different fat contents can be pickle cured successfully, but those with the highest fat content make the best products. The quality and type of the fish raw material and salt is important. Fresh raw material is better. Salt should be of reasonably small grain size to facilitate close contact with the fish surfaces and rapid dissolution, but not so fine as to impede drainage of expelled flesh juices. Salt containing more than traces of iron (10 ppm) or copper (1 ppm) gives rise to unsightly yellowish or brownish colour in finished white fish products and hence it should be avoided. However, it should contain about 0.5% calcium and magnesium (as sulphates), because these metals impart a desirable whiteness and rigidity; higher concentrations are undesirable, because they cause excessive bitterness and brittleness. White fish cured with pure sodium chloride tend to be flexible and amber in colour.

The aim of salting is to obtain an even distribution of salt through the fish and therefore uniform sizes of fish should be stacked together or placed in the same region of the barrel or pickle container; also more solid salt should be placed near the thickest part of the fish. Air spaces should be avoided in order to reduce the risk of rancidity development. In this connection, barrels or wooden containers used for holding fish and salt should be as air tight as possible, the entry of air can cause the development of rancidity. Drying should be carried out as swiftly as possible to reduce the risk of spoilage. The humidity of the air needs to be low and the temperature as high as possible consistent with the avoidance of cooking and early case hardening. Good control and rapidity of drying in humid, cool climates can be obtained by suspending the salted fish in a chamber through which warm air is passed. Salted fish can be dried out of doors in dry, warm climates, but protection against rain, insects, bird and strong direct sunlight is necessary. Even and more rapid drying throughout individual fish and between the fish in a batch is aided by re - stacking ('press- piling') for several days after some drying has taken place. This has the effect of equilibrating the moisture content throughout the mass; afterwards drying is resumed.

The deterioration occurring during processing of salted fish is mainly microbiological. Because the anaerobic conditions in pickle or wet curing of pelagic species rarely support the growth of microorganisms and hence, these deteriorations are almost associated entirely with dry or stack curing of white fish. In uncommon types of pickling of whole herring using mixtures of salt and sucrose, the process is occasionally troubled by the occurrence of ' ropy brine', a slimy condition caused by the bacterial formation of a polysaccharide from the sugar. But, this can be avoided by keeping the pickle chilled. There are four types of microbiological deterioration ims salted white fish, such as slime, putty fish, pink and dun. During conditions of high ambient temperatures, high humidity or inadequate penetration of salt, certain groups of bacteria capable of living in low concentrations of salt (6 -12%) may multiply and attack the fish, resulting in a sticky, 'slimy' coating and off-odour. The remedy is to reduce the temperature and humidity and to accelerate drying of the outer layers by increasing ventilation and to enhance penetration of salt. The deterioration 'putty' occurs in the thick parts of the fish, where the rate of increase in salt concentration is slowest.

Degradative bacteria may gain a hold, if the temperature is too high, and reduce the fish to the consistency of putty. 'Pink' takes the form of spots or areas of pink or reddish colouration on the outside of the stack. If left untreated, the condition spreads to cover and partially penetrates the whole stack, turning the fish into soft and bad smelling. It occurs only when the salt concentration is high (over 10 -15%) and is caused by the growth of certain groups of pink or reddish coloured bacteria (halophiles) such as *Halobacterium* sp., *Halococcus* sp., *Sarcina marscenes* and *Pseudomonas salinaria*. Halophiles originate in some kinds of salt and assuming the premises are not contaminated, the occurrence of the deterioration can be prevented or reduced by changing the salt more or less free of the offending bacteria, by heating the salt before use, or since the contamination tends to decrease with time, by using salt that has been stored for a year or so. Premises and equipment can harbour halophiles and, if an outbreak of 'pink' occurs, a thorough cleaning programme should be instituted using plentiful amounts of water, preferably containing disinfectant suitable for food use. Fumigation with sulphur dioxide is also useful.

'Dun' caused by halophilic or halotolerant fungi such as *Wallemia* or *Oospora*. It is the term used to describe a peppering of light brown or light yellowish brown spots, particularly noticeable on the cut surfaces. *Sporendonema epizoem*, which is able to grow at salt concentrations of up to 10–15%, is more often seen on light cures. The best preventive measure is good housekeeping of the premises and surroundings by removing rotting vegetation and by having a regular cleaning regime. Dry, well-ventilated and painted buildings should be employed, if possible. Sorbic acid, where it is permitted by food regulations, is effective at 1.0% level against this kind of mould; and it is applied by dipping the fish in an aqueous solution before stacking. The storage life of salted fish depends upon the salt concentration, the moisture content and the prevailing atmospheric conditions. The lower the salt and the higher the moisture content, the shorter is the shelf life. Storage deteriorations in salted white fish take the form of a gradual softening and development of off-odours predominantly due to enzyme action; if the moisture content rises to too high a level, the fish go badly as a result of bacterial spoilage. Effective packaging can prevent the spoilage to a considerable degree consequent on the absorption of moisture from ambient air of high relative humidity. The best materials for this purpose are flexible plastic films, plastic containers or bitumen lined brown paper. Salted fish is often prepared in countries with warm humid climates, where insect infestation or rodent attack are prevalent and cause extensive losses. In salting of fish, it is important to take into consideration of the factors such as purity of salt, amount of salt, duration of salting and weather conditions. Common defects associated with salted fish are pink or red and dun or black spots, rust and maggot infestation.

1. *Pink or red and dun or black spots*

A large quantity of cured fish is spoiled by the attack of fungus and red halophiles. These products develop discolourations as pink patches which later on result in the porous structure of the product. High moisture content of the product favours microbial growth, while red attack is due to the high salt content. These can be prevented by dusting the dried product at the time of packaging with a mixture of 3 parts of sodium propionate and 97 parts of sodium chloride at the rate of 10% by weight or sprinkling of 0.1% calcium propionate on cured fish prior to packing prevents the growth of fungus and red halophiles.

2. *Rust*

In this type of spoilage, the surface of the salted fish acquires the colour of rusted iron and the product gives unpleasant taste and rancid odour. Rust is due to the oxidation of fat in the presence of atmospheric oxygen. It has been found that the traces of calcium and magnesium enhanced the oxidation process. The best method of controlling this defect is to prevent the contact of salted fish with air. Fish should be kept covered with brine during salting. Salted fish, after drying, should be properly packed and pressed down in the container.

3. *Maggot infestation*

This is a common defect in salted fish. The cheese fly deposits its eggs on the salted fish or on the sides of the container and when the cycle of development is completed, the maggots come out and infest the whole lot of fish. The best preventive measures are to keep the premises clean, to have concrete floors and to ensure that the salted fish is properly covered so as to avoid the entrance of flies.

Sodium benzoate has shown promise in countering halophilic bacteria in salted fish, while dipping in 1.0% sorbic acid as potassium sorbate solution inhibits 'dun' mould. Insecticides used for direct application to dried fish must be sufficiently toxic to the target insect pests, while not leaving residues harmful to the consumers. The only currently recommended maximum residual limits (MRL) by the FAO/WHO Joint Meeting for Pesticide Residues (JMPR) and the Codex Alimentarius Commission on fish for pyrethrins is 3 mg / kg and piperonyl butoxide is 20 mg / kg. The recommended insecticidal treatment is to dip in a solution of 0.125% pyrethrins or 0.25% piperonyl butoxide. Pirimiphos - methyl and chlorpyrifos - methyl might be acceptable as fabric sprays, where dried fish is handled or stored. An application rate of 0.5 - 1.0 g / m^2 of insecticide may be effective. The fumigants, which are in general use, are methyl bromide and phosphine. Methyl bromide is applied under partial vacuum at a dosage rate of 80 g / m^3 for 2 h or under normal atmospheric conditions for 6 h. Phosphine is applied at a dosage of 0.1 mg / lit for 72 h or 0.2 to 0.5 g / 50 kg fish for 2 to 3 days. Fumigating the fish with sulfur gas in an enclosed space can also control insect pests in sun-dried fish. The fish are arranged at a distance of 1 to 1.5 m from the floor and fuming done for 1½ days. The sulfur is used at the rate of 50 g /m^3 of storage space. Fuming treatment has no deleterious effect on the appearance or quality of the product. The drier the product and the lower its salt concentration, the better the conditions for the larvae of the beetle to thrive and spoil the fish, often from the inside without damaging the skin colour. After the fumigating treatment, the fish are shaken and ventilated or arranged on a platform. Under the effect of the sun's rays, the larvae come out of the fish, which are then collected and destroyed.

Drying is a process by which water from a moist substance is removed. If air is used to carry away the water vapour formed, the process is called as airdrying. Two kinds of drying are natural drying and artificial drying. The preservation by drying depends on reducing the moisture content to a level where microorganisms and most intrinsic enzymes become inactive. Thus, the amount of drying has to be considerably greater than that required for salted products. Generally moisture content of not more than 15 - 20% is aimed at, this being the upper limit below which moulds do not grow. The aim in production is to reduce the moisture content quickly enough to avoid the accompanying spoilage. Control of quality changes during natural drying is obviously heavily dependent on a steady, reproducible climate. Drying tunnels of various designs are available with which it is possible to easily produce dried fish of consistent quality. The best method of retaining natural quality in dried fish is freeze drying, but this very expensive procedure has so far only been applied to costly fish products like shrimps. Raw material should be as fresh as possible and wholesome. However, when very fresh squid are dried too rapidly, black pigment cells in the surface layers of the body are liable to become prominent causing an unwanted dark appearance in the finished product. This can be overcome by soaking the dead animals in freshwater, which causes the pigment cells to contract.

Drying is expedited by creating as large a surface area as possible by splitting large fish into thin sections; breaking or mincing; spreading; hanging or laying out thinly; turning over laidout fish frequently. Under given conditions of temperature, humidity and airflow, moisture escapes from the surface of the fish at an approximately steady rate. Making the surface area as great as possible maximizes evaporation and minimizes drying time. Generally, initial removal of some body juices by pressing out or by heating until the flesh proteins coagulate is particularly advantageous. The water in raw fish is held quite firmly and is accordingly removed very slowly. Coagulation releases the water as a separate juice, which then evaporates more rapidly. Cooking before drying also has the advantages that it normally avoids the occurrence of case hardening during rapid drying and arrests the actions of microorganisms and enzymes. When drying naturally, protection against direct strong sunlight is often necessary in order to avoid partial cooking and break up of the flesh or case hardening. Protection against rainfall is also essential. If the surface of drying squid becomes wet through exposure to rain or high humidity, it develops an undesirable reddish colour caused by the leaching out of pigments lying just under the skin. Where humidity or insects increase at night, the fish should be taken indoors. Fish spread outdoors should be protected against fouling by and depredations of birds and other animals.

Properly prepared and packed dried fish product should keep for several years at average ambient temperatures. In humid conditions, where the relative humidity is greater than 75%, unprotected dried fish of initially low moisture content take up sufficient water within a few weeks to allow mould growth. A major problem in the control of quality of dried fish is insect attack. Good packaging is one answer. Dipping before drying of fish in a dilute solution at 0.125% level of the odourless insecticide pyrethrin, or light dusting of the finished product with the dried substance suitably diluted, offer good protection against insect attack. Where it is feasible, fumigation of dry fish storage spaces with methyl bromide is effective in keeping insect infestation under control. Very slow deteriorations in non-fatty fish are the enzymatic production of off-flavours and chemical reactions such as maillard reactions between carbonyls and amino compounds that result in yellowish or brownish discolourations and burnt off-flavours. It is very important to control these deteriorations, which occur in freeze dried prawns. Maillard reactions may also occur between pentose sugars obtained from nucleic acids (DNA, RNA) or aldehydes obtained from oxidative rancidity of products and amino compounds present in the fish. Unless protected, dried fatty fish quickly becomes rancid and softens. Packing in an inert gas or vacuum offers good protection or effective dispersion of antioxidants through the product can considerably lengthen the storage life. Common defects in dried fish are high sand content, loss in nutritional value, loss in other quality.

1. High sand content in dried spratts (Anchoviella sp.)

The best way to avoid the presence of high sand content is to dry the fish without contact with sand. But, under the existing conditions, it may be difficult to adopt such methods. In that case, a mechanized fish cleaner can be used for removing sand from the conventionally dried product.

2. *Losses in nutritional value*

The exposure of wet fish to high temperatures is the most likely cause of protein damage. Drying at 75°C and smoking at 100°C lower lysine availability and net protein utilization (NPU). There is no evidence that the temperatures typically reached during sun drying cause any appreciable loss of nutritional value of the protein. But, excessive heat treatment is known to impair the nutritional value of fish proteins, as a result of a variety of chemical reactions. It has been found that sometimes oxidized fish fat is known to react with the proteins and damage the nutritional value of protein. The nutritional value of the oxidized fish fat is appreciably lower than that of the fat in the natural form. There is also some evidence that lipid peroxides, an intermediate product of oxidation, are toxic. Vitamins A and E can be destroyed by lipid peroxides. It has been observed that when the fish is dried, the fish fat has little protection from oxidation. The drier the fish, the greater is the tendency for the lipids to get oxidized.

3. *Losses in other quality*

Deterioration starts immediately after death, and losses can occur until the a_w is reduced sufficiently to prevent the normal spoilage processes. Enzymes, particularly those in the gut, start to digest fish tissues. Certain moulds and bacteria can grow on moist products giving the fish an unattractive appearance and odour. Minor growth is often scraped off and the fish redried, but severe attacks will soften the flesh causing it to break up and losses can eventually be total. Moist fish is susceptible to damage by blowflies and their larvae. Blowfly larvae consume the fish flesh, and only when the fish has been sufficiently dried, it becomes unattractive to the adult fly for egg laying. Drying fish is often particularly exposed to birds, rodents and domestic animals. Under very humid conditions, the dried fish will reabsorb moisture, which makes them susceptible to losses. A relative humidity of over 70% is conducive to mould attack. Products, which are sufficiently well preserved to prevent microbial attack, are still susceptible to insect attack and may also be attacked by scavenging animals. Beetles, especially *Dermestes* spp. feed on dried fish, completely consuming the flesh. Mites are also occasionally found in dried fish. Losses can also result from the physical disintegration of the fish.

4.5. OTHER FISH PRODUCTS

Marinades are the products that are preserved by the combined action of dilute acetic acid and salt. The inhibitory effects on bacteria and enzymes are greater at higher concentrations, but, since marinades are consumed without any further preparation, shelf life is limited by the upper concentration, particularly of acid, that is palatable. Marinading is normally a two stage process with the transfer of the fish in between. The aim in the first stage is to render the fish, normally filleted, rapidly as sterile as possible, while at the same time developing the characteristic basic texture and flavour; for this purpose immersion for atleast a week and frequently much longer in concentrated pickle solution containing 5 - 10% acetic acid and 10 - 15% salt is employed. In this process, the protein of the flesh is coagulated and the remaining small bones are softened. The aim in the second stage is to maintain a palatable level of preservative that keep the product for a reasonable length of time. A final product in 1 - 2% acid and 2 - 4% salt keep in good condition for atleast 3 months at near 0°C. The eating quality of this product is very much dependent upon freshness and lack of damage or contamination of the raw material. Frozen and thawed fish of good quality is quite acceptable. Fatty species and herring make the best traditional marinades; a minimum fat content of about 10% is required. Very high fatty fish should be avoided, because an unpleasant looking and possibly rancid layer of fat may in time rise to the top of the second covering pickle. Salt cured fish are also suitable after some of the impregnating salt has been removed by soaking in water. As the appearance of marinades is highly important, the fish should be cut, cleaned and washed in brine very carefully to remove traces of viscera, blood, slime and dirt. Boiled or fried fish can also be marinaded.

The whole fish or fillets should be mixed thoroughly with the first pickle introducing them individually and stirring or agitating the mass from time to time to ensure uniform and rapid penetration. Apart from effects on flavour and keeping quality, the texture of the final product is dependent on the concentrations of preservatives used; more acid softens, an effect that is moderated by increasing salt concentrations. Thus, the relative proportions of acid, salt and fish in both pickles is of great importance and should be strictly controlled for each type of product. Fish, that on removal from the first pickle, are obviously spoiled or show discolourations cannot be improved by transfer to the second pickle and should be discarded. Special attention should also be given to the quality of other ingredients like spices and vegetables, which are added to the final pickle. Marinades are often packed without heat treatment in transparent jars or in cans suitable for retail display; all containers of this kind should be washed thoroughly and inspected before use for damage or faulty closures. Fish should be trimmed and packed neatly to give better appearance.

Some bacteria and proteolytic enzymes are weakly active in the second pickle and the product gradually develops off-odours, discolourations and softening at a rate depending on the temperature of the storage. Marinades cannot be frozen because break-up of the flesh can occur. Addition of small amounts of the bactericidal substance such as hexamethylenetetramine to the final pickle prolongs shelf life, but is not permitted in many countries.

Other heat processed fish includes the products such as sausage, ham and kamaboko, which are boiled, fried or steamed but not packed in hermetically sealed containers. Depending upon its intensity, heat processing kills a varying proportion of bacteria and destroys most flesh enzymes. Thus, spoilage results from the heat-resistant microorganisms that have survived the cooking process as well as from those that have subsequently contaminated the product during handling and packing. The onset of spoilage is always delayed in comparison to corresponding fish that has not been heat processed. Some food poisoning microorganisms are rather heat-resistant and others are picked up by handling after cooking and may prove a hazard to health before the changes that result in obvious spoilage indicating that fish is not acceptable for human consumption. Therefore, the precautions that must be taken to ensure with a high degree of certainty that heat processed fish is microbiologically safe are more stringent than those appropriate to raw fish. Cooking or pasteurization should be carried out strictly according to procedures worked out and proved to be adequate, and more attention given to hygiene and cleaning both before and after heating. Because of the danger of cross- contamination of cooked warm fish with spoilage bacteria originating from uncooked fish, the processing of both the fish should be kept well separated.

Cooked fish should not be re-heated under conditions that encourage growth of food poisoning microorganisms. Cooked fish quickly loses moisture when hot and becomes unpleasantly dried out. It is not a good practice to keep the cooked fish in a hot cabinet for more than 15 min. before further preparation. Cooked fish, which is cooled slowly or kept warm for long periods, can also develop a rancid or cardboard-like odour and flavour. It should be cooled rapidly by immersion in clean potable cold water. Generally, heating offers effective means of arresting spoilage. Shrimps are often cooked before further processing, which allows the primary treatment to be carried out on fishing vessels. Cooking onboard immediately after capture helps to preserve flavour and colour, but because of the risk of subsequent bacterial contamination, the product should be handled as quickly as possible. Seawater used to cool shrimps after cooking onboard should be clean and uncontaminated. Another method applicable to crab and shrimp meat is to heat them in suitably heat-resistant plastic pouches so that the centre is kept for 5-10 min. at the pasteurizing temperatures (80 -85°C). Most of the organisms are killed, and since the pouch offers good protection against re-entry of microbes, a much enhanced shelf life of 4 - 6 weeks at near 0°C is possible before the product becomes inedible.

A common preservative technique in the Far East countries is to boil whole eviscerated fish in strong brine, to cool and distribute it without further handling. Shelf life of 2 -10 days, depending on ambient temperature, is attainable and these can be extended by re-boiling after some days. Cooking fish in pre- or in-rigor may present quality problems. Flesh may break up and become very tough or rubbery. Flavour is also often peculiarly metallic or watery and quite abnormal. Holding the fish for a few hours at chill temperatures, until rigor has passed, removes this quality defect. The traditional preservatives including salt, vinegar and acetic acid, alcohol and natural smoke are universally acceptable, though even the natural smoke has come under scrutiny on account of possible carcinogenic effects of some of its trace constituents. Very few artificial preservatives such as sulphur dioxide, sorbic acid, benzoic acid and hexamethylenetetramine are now permitted in few countries for the use in fish products. It is always better to use the permitted flavouring and colouring agents in fishery products.

Common defects associated with fish sausage are slime formation, softening, spot formation, gas formation, acidification, liquefaction and discolouration. Slime formation is caused by *Streptococcus, Leuconostoc* and *Micrococcus. Bacillus pantothenticus, B. circulans, B. coagulans, B. subtilis* and *B. putrefaciens* are responsible for softening of sausage and spoilage. Black spot formation is caused by *B. coagulans. Lactobacillus* and *Clostridium* produce gases such as NH_3 and CO_2 in sausage. Acidification, which is similar to flat sour spoilage in canned fish, is brought out by *Lactobacillus. Streptococcus* is the causative agent for liquefaction and discolouration. *Bacillus* is the prime spoiler in surimi. Volatile fatty acids such as lactic acid and acetic acid and alcohols such as ethanol and butanediol are formed through sucrose and sorbitol metabolism by *Bacillus* during spoilage of surimi crab legs. Production of volatile acids and bases and biogenic amines such as cadaverine and putrescine are decomposition indicators for vacuum-packed surimi. Pathogens such as *Listeria monocytogenes, Clostridium botulinum, Aeromonas hydrophila* and *Yersinia enterocolitica* encounter in the surimi based aquatic foods.

CHAPTER 5

METHODS OF ASSESSING QUALITY OF AQUATIC FOOD

Many methods have been proposed for measuring aquatic food quality. Few have been shown to have shortcomings like inaccuracy or impracticability serious enough for them to be rejected. Others are suitable only for research or product development. The methods of assessing aquatic food quality can be conveniently classified into three groups such as sensory, physical biochemical and microbiological methods. In the early days, when quality control was first introduced, importance was only given to sensory evaluation to assess the quality of food product. But in the recent past, instrumental techniques have been developed for measuring food quality and statistical methods have also been introduced for interpreting the results.

There are advantages and disadvantages of subjective and objective methods of assessing the quality of aquatic food. The sensory evaluation of the quality of food or judging the quality of food with the aid of human senses, where the feelings of liking, pleasure, acceptance, valuation or prejudice are allowed fully or the deciding factors, where the results depend purely on the individual and personal; then they are called subjective method of evaluation. They are also called hedonic, because the method evaluates the degree of pleasure experienced by the person evaluating the quality. The physical or chemical evaluation of quality by using instruments, where the individual human element does not play a role, is known as objective method of evaluation. It can also be referred to as defective evaluation. Since the human instrument is used in subjective method, it suffers from the human limitations of being influenced by environmental conditions, state of health of the individual, lack of absolute point of reference, which may cause an error, tendency for comparative rather than absolute evaluations and finally personal bias, which may enter the process of evaluation consciously or subconsciously. The method of objective or defective evaluation is free of the human element and this may give a true picture of the quality of the product, but here again human evaluation has to be done, because ultimately consumers judge the product by his/her senses not by the instruments. Further, the method of objective or defective evaluation is a time consuming process and it is also not possible to evaluate the quality of the food by the defective method without first damaging it. The characteristics like colour and texture can also be measured by instruments. The chemical measurements aim at determining the quantity of the degradation products that have been produced as a result of spoilage and a measure of these indicates the extent of spoilage or the degree of freshness of the material. The subjective method of evaluation does not require a well-equipped laboratory. The evaluation can be done even as the fish is displayed for sale in the market, whereas, the objective or defective method of evaluation requires a well equipped laboratory. Trained personnel are not required for subjective assessment of evaluating quality. Sometimes little amount of training required can be easily imparted or by experience, the personnel can acquire expertise to evaluate quality in a short time. However, an adequate and exhaustive training is required in the objective or defective method of evaluating quality. As sensory evaluation does not require any laboratory facilities, it may be more economical. Sensory evaluation is a necessary one, because instruments are not developed or being developed are not accurate or reliable or convenience or quick to test the quality.

5.1. SENSORY METHODS

Sensory evaluation has been defined as a scientific discipline used to evoke, measure, analyze and interpret reactions to those characteristics of foods and materials as they are perceived by the senses of sight, smell, taste, touch and hearing. Sensory evaluation tests can be divided into two general categories: affective or subjective and analytical or objective. The purpose of affective tests is to evaluate preference and / or acceptance of products. In affective tests, the spontaneous, personal reaction of the panelist is evoked in order to determine preference or acceptance. These are subjective tests designed to capture the original and spontaneous impressions of the panelists. Analytical tests evaluate differences or similarities, quality, and / or quantity of sensory characteristics of products. In analytical tests, some aspect of sensory quality of the product itself is of interest and not the personal reaction of the panelist. The panel is the analytical tool. The International Organization for Standardization (ISO) defines an objective method as "a method in which the effects of personal influence are minimized". Procedures have been developed for analytical sensory tests in an effort to control or minimize the effect that psychological and physical conditions can have on the panelists' reactions. When these procedures are followed, analytical sensory tests meet the requirements to be called "objective". Sensory or organoleptic methods are fully dependent upon the human senses for assessing the quality. All the senses except hearing are used throughout the fish industries to judge quality such as sight (appearance), touch (texture), odour and flavour (smell and taste). Since the consumer uses only his/ her senses in deciding what he/ she likes, the sensory method of evaluating fish quality is the best method, which offers the best opportunity of getting a valid idea of what the consumer wants. Sensory methods also have the great advantage that human beings are very adaptable and can switch easily from, for example, testing odours to visual inspection for defects. Furthermore, for some tasks human senses are better at recognizing complexities and are more discriminatory than instruments. Their main disadvantages are that their responses can vary, particularly with fatigue or outside distraction, and that using people can be expensive and inconvenient.

I. Quality factors assessed

Quality factors assessed by sensory methods are sight and touch and odour and flavour.

a. Sight and touch

The selection for species and size falls under this heading. It requires a minimum of training to get fishermen to segregate the catch into species by appearance that are worth keeping and into the size groups that are wanted. Sensing of weight is also brought into play when sorting by hand for size. The detection of deteriorations and defects in almost all the cases is very efficiently and rapidly accomplished by sight. It would be very difficult to make a machine capable of detecting poorly trimmed fillets, or fillets possessing too large an area of skin. Testing with fingers and eyes remains the only effective way for detecting the bones in the fish flesh. The matching of colour can be done very effectively by eye. The sense of touch in fingers or mouth are used in assessing textural attributes such as firmness, softness, mushiness, rubberiness, woodiness, mealiness, succulence and dryness. If a tasting test on cooked fish is conducted, it is convenient to include the assessment of texture as well as of odour and flavour in the sensory evaluation.

b. Odour and flavour

Flavour is usually meant to include much of what is experienced on smelling through nose and tasting through tongue. Hence, odour and flavour can be taken together. Anyone can distinguish between the smell of fresh and spoiled fish. With some practice, the whole pattern of changes in odour between very fresh and very spoiled can be easily and rapidly differentiated so enabling to accurately determine the degree of freshness. Similarly, off-odours, taints and unusual intrinsic odours are readily detected and their intensity judged reproducibly. Tasting is also the most reproducible method, which is capable of assessing the changes in character of the flavour far more satisfactorily. Tongue can detect four basic tastes such as sweet, salt, acid and bitter. Different levels of these tastes can be easily distinguished so long as comparisons are possible, but the measurement with accuracy of degree of saltiness or acidity in an absolute sense is difficult.

II. Types of sensory assessment

Depending upon how they are used, two types of assessment can be distinguished. The first is the dispassionate, unbiased, descriptive assessment of individual or groups of quality factors. Examples of responses in this kind of assessment are: this fish is fresh/ salty/ stale/ sour/ tough/ sandy/ has too many bones. Sometimes this kind of sensory method is called objective, because the person carrying it out tries to remove from his/ her judgement all feelings of liking or disliking. The second is where, after examining the product, feelings of liking, pleasure, acceptance, valuation or prejudice are fully allowed, as for example: I think this fish is excellent/ inedible/ unacceptable. Judgements of this kind are termed on occasions either subjective, because they are entirely personal, or hedonic because they relate to pleasure or degrees of it. The utility of subjective sensory methods in practical quality control is very limited because the result does not help much in deciding what corrective action to take if something is wrong. Hedonic judgements can be used to define initially the quality of products that consumers like and which in production and marketing should be aimed at. The descriptive, objective method in which error is reduced to a minimum is of much more use. If necessary, the attributes of quality that consumers want can be related to the results.

a. Scales, scores and grades

It is not always enough to simply say that whether the product is good or bad in the evaluation of fish quality, because there may be different gradations of quality, defects or deterioration. Wherever a graded series of changes has to be contented with, it is useful to the person carrying out the assessment to construct a scale showing exactly how the changes occur. A scale provides the assessor with a fixed yardstick which he/ she can use on different occasions. Scale consists of a series of steps, each being described in a distinct form of words. About 4 or 5 steps would normally form a scale. It is always possible to construct a scale showing change or incidence of the use of value words such as light, trace, medium, moderate and very highly. A well-known scale showing stages by which the odour changes in spoiling white fish is given below; the description at the top relates to absolutely fresh fish, the others to decrease in freshness to absolutely putrid at the bottom.

Fresh seaweedy

Loss of fresh seaweediness, shellfishy

No odours, neutral

Slightly musty, mousey, milky, caprylic

Bready, malty, beery, yeasty

Lactic acid, sour milk, oily

Acetic or butyric acid, grassy, slightly sweet, fruity

Stale cabbage, turnipy, wet matches, phosphene- like

Amine, tyre-like

Hydrogen sulphide, strongly ammoniacal

Indole, faecal, nauseating, putrid

A further development of scales is to denote the different steps by numbered "scores" from 0 or 1 upwards; the spoilage example just given above can be made into a 10 point scoring scale. This is not only a convenient shorthand device for identifying the steps, but allows the results from different sensory assessments to be pooled, averaged and so on by the process of ordinary arithmetic and statistics. If the numerical sensory spoilage 'scores' of few fish taken representatively from a batch are averaged, the result can be said to describe the freshness quality of the batch. Depending upon where the acceptable quality level is set, the batch may then be accepted or rejected according to where its average freshness falls. Since defects and degrees of deterioration are undesirable, it is usual to think of their increase as warranting the award of demerit points or scores. i.e. the highest numbers of demerit points would obviously be worst. Where a number of defects or deteriorations occur together in the same product, it is a common practice to sum the scores or points allocated to the individual attributes to give an overall demerit score or grade. It is necessary when setting the scoring or points level above or below which the product is unacceptable, to allow a tolerance, because a series of fish products without defects is an impossibility. Grades have essentially the same meaning as scores, but in most usage tend to be simpler, less finely subdivided or combinations of several quality attributes considered together. A grade 1 or grade A product would possesses few defects or signs of deterioration of any kind; a lower grade, several of one or more kinds. Grades are often defined in terms of the total number of defects or demerit points. An example of a freshness-grading scheme for whole, chilled lean finfish based partly on the freshness odour scale already given is given in Table 1.

Table 1. Freshness Grading Scheme for whole chilled lean finfish

Parts of Fish	Extra	A	B	C (unfit)
Skin	Bright, shining, iridescent, opalescent, no bleaching	Waxy, slight loss of bloom, slight bleaching	Dull, some bleaching	Dull, gritty, marked bleaching and shrinkage
Outer slime	Transparent or water white	Milky	Yellowish grey, some clotting	Yellowish brown, very clotted and thick
Eyes	Convex, black pupil, translucent cornea	Plane, slightly, opaque pupil, slightly opalescent cornea	Slightly concave, grey pupil, opaque cornea	Completely sunken, grey pupil, opaque discoloured cornea
Gills	Bright red, mucus translucent	Pink, mucus slightly opaque	Grey, bleached, mucus opaque and thick	Brown, bleached, mucus yellowish grey and clotted
Peritoneum	Glossy, brilliant, difficult to tear from flesh	Slightly dull, difficult to tear from flesh	Gritty, fairly easy to tear from flesh	Gritty, easily torn from flesh
Gill and internal odours	Fresh, strong seaweedy, shellfishy	No odour, neutral odour, trace of musty, mousy, etc.	Definite musty, mousy, bready, malty, etc.	Acetic, fruity, amines, sulphide, faecal

Such grading schemes are useful whenever the quality of batches of fish has to be rapidly assessed, such as port markets or factory reception areas. The quality control inspector scans the fish taking into account more or less simultaneously a number of different attributes before allocating a grade.

b. Judges or panelists

A judge or panelist is any person, who carries out sensory assessments. Most judges should be experienced or trained to assess quality objectively. Adequate control of quality is done in most of the fish industries by a single expert judge. But, by having two or more trained or experienced judges, who normally act independently but assess the same product, the risk of major mistakes through prejudice or bias is almost entirely eliminated. If all the judges assess the same sample or different samples from a batch, using the same points scoring system, the averaged results give a correct measure of the attribute in question than is possible with a single judge. The maximum number of judges that are required in this type of work is about six. A group of judges is often called a taste panel, particularly when their judgements involve tasting.

c. Precautions

Sensory judgements are sometimes influenced by external factors. The colour of a product appears different under different types of lighting. Therefore for carrying out colour comparisons, standardized lighting conditions should be used. The detection of a special odour in a working environment like a market can be hampered by extraneous odours; practice is therefore usually needed so that the unwanted interference can be ignored. If samples are tasted after heating or cooking, the process should be carried out in an identical fashion on every occasion. Sensory assessments require a certain amount of concentration and so distractions should be avoided. Ideally tests should be conducted in an area or separate room isolated from processing and other industrial operations.

5.2. PHYSICAL/MECHANICAL/INSTRUMENTAL METHODS

A few mechanical devices have been developed for grading the fish and shellfish according to sizes. Machines are also available for separating males and females of some fishes. A number of instruments for measuring moisture and fat are also available. A device for moisture consists of an infrared heater to quickly dry the finely comminuted sample, which is weighed simultaneously. The minced sample is rapidly extracted with a fixed quantity of a fat solvent for fat determination. After intensively shaking them together, the solvent is drained off and its specific gravity measured automatically with a special balance. The specific gravity is proportional to the amount of fat in the solvent. Instruments for detecting or measuring colour, parasites or contaminants have either not proved practicable or given problems of commercial application. A few physical methods are available to measure the deteriorations and defects in raw material and products. Progressive and marked changes in the electrical properties of the skin and underlying tissue provides a means of measuring the degree of spoilage in most types of chilled whole fish. Three different direct reading instantly responding instruments based on this principle have been developed. They are Torrymeter (TM) of UK, Intelectron Fish Tester (IFT) of Germany, and RT- Freshness Grader (RTFG) of Iceland. TM has readings from 0 to 16, while IFT has a wider range from 0 to 100. RTFG has been reported to be efficient for on the spot testing of freshness in the fish landing centres.

The testers provide a meter reading that can be calibrated in terms of the degree of spoilage or the shelf life remaining before the fish is inedible. It is small enough to be carried around in the hand for direct application to the surface of the fish. No damage is created to the sample by the test. Satisfactory readings cannot be obtained with frozen and thawed fish, fillets or flesh. However, a reliable, robust and accurate instrument of this general type could provide a very useful standard tool for checking and monitoring the freshness quality, and it is believed that the British design fulfills these requirements. Its advantages are that it is independent of fallible human judgement, does not require highly trained or experienced personnel, never tires and in certain circumstances can test more fish in the same time than is possible by sensory methods. Its main disadvantage is that readings on single fish are by no means a clear guide to freshness as judged sensorily and for this reason it is essential to test several fish in order to obtain a good measure of the average freshness of a batch. The scores shown by this instrument should be standardized for different species and for same species at different geographic locations due to the possible difference in the proximate composition of fish.

A number of instruments for measuring firmness in frozen, chilled and canned fish have been designed. Two basic types of firmness instruments are those that measure the force necessary to press a plunger or set of sharp needles into the sample and one that measures the degree of fragmentation when fish flesh is subjected to a fixed amount of disintegration in a homogenizing apparatus. The former, texturometer/ universal testing machine (UTM), is in use in some salmon canneries. In the latter, designed for frozen products, the degree of fragmentation, measured by an optical method, is less in firmer specimens and thus provides a measure of textural eating quality; and such test is called cell fragility test. Sometimes, tristimulus colorimetry is used to measure the colour of the product. The pH of fish flesh is directly related to firmness. A sample of low pH can be rated tough by both sensory and instrumental methods. The direct measurement of pH on raw material before acceptance for processing has been proposed as a means of avoiding fish that could be too tough or otherwise of undesirable pH. The measurement requires a reading of the electrical potential between glass electrode and a calomel electrode, both directly inserted into the flesh. Determinations of drained weight, net content or 'drip' require only simple apparatus such as standard sieve, balance and clock. Small differences in the refractive indices of the oils from different species allow a tentative identification of an unknown sample of fish to be made. The control of processing time and temperature is of vital importance in ensuring the maintenance of quality, and therefore the clock and thermometer are the two most important physical instruments. "Bionic Nose" developed by the University of Nagasaki, Japan has been used for testing odours.

5.3. BIOCHEMICAL/ CHEMICAL METHODS

They are used for determining the composition of raw material and products as well as for detecting the deteriorations in the products.

a. *Composition*

Determination of moisture content may be necessary, where the presence of excessive water in fish raw material is suspected or where it is important to dry a product to specified water content. In such conditions, a representative sample is weighed out accurately and dried in a hot air oven or vacuum oven at a temperature sufficient to drive off the water in a particular time. After a prescribed period, the sample is re-weighed and then the moisture content is calculated. Determination of protein content is required in connection with the measurement of the fish content of a product such as fish cakes, fish fingers, etc. The method depends upon knowing the protein content of the fish used in the product and if necessary, making a correction for the protein contained in other ingredients. Protein content is assessed by determining the nitrogen content of the sample and multiplying it by a factor, 6.25, representing the inverse of the known nitrogen content of the protein. Nitrogen is commonly determined by kjeldahl method. Biuret method may also be used. Chemical methods for fat are preferred. The weighed sample is extracted with a fat solvent in a special refluxing apparatus such as soxhlet apparatus and the solvent evaporated from the resulting solution of fat leaving a dry residue of the latter, which is then weighed. Bligh and Dyer extraction method may also be used.

Measurement of mineral content is occasionally important, as it is a good measure of bone, shell or sand content of a product. The mineral content is determined by burning off the organic part of a known quantity of the product at a high temperature in a muffle furnace and weighing the residue of ash. Determination of salt as sodium chloride is frequently necessary as it is an important constituent of many fish products. It is best carried out by dissolving the whole sample in nitric acid and measuring the concentration of chloride in the solution by titration with silver nitrate solution. In acid-preserved fish like marinades, it is necessary to determine its acid content. It is achieved by titrating a sample with standard solution of alkali after blending in a known volume of water. Analytical methods for determining the metal contents such as mercury, lead, cadmium, zinc, etc., chlorinated hydrocarbons, radioactive isotopes, colouring matters, additives and preservatives are complex and are normally only undertaken by well equipped laboratories. A special compositional method, which is of more widespread importance, relates to the identity of fish species in products. In such cases, the recommended method is to extract the proteins of the fish into an aqueous solution and then to separate them in gel electrophoresis, which provides a pattern of bands of different intensities. Every species of fish has a uniquely different pattern that serves unequivocally to characterize it. By this method, the unknown samples can be identified by matching their pattern against those of known species of fish.

b. Deterioration

The chemical and biochemical methods mostly measure the extent of spoilage in the chilled fish and of oxidative rancidity. Three well developed methods for measuring spoilage in the chilled fish are available, which depend upon the complex series of changes in flesh constituents brought about by autolytic enzymes and putrefactive microorganisms. With certain precautions, these methods can also be applied to products such as frozen, dried and canned fish in order to provide a measure of the amount of spoilage that has occurred before processing. It should be noted that they do not necessarily provide any indication about wholesomeness. The important classes of spoilage microorganisms found in tropical fish are *Pseudomonas, Alteromonas, Flavobacterium, Acinetobacter, Aeromonas* and *Moraxella*. The spoilage bacteria are characterized by their ability to produce H_2S, reduce trimethylamine oxide (TMAO) to trimethylamine (TMA) and convert urea to ammonia. Many volatile sulphur compounds are also produced by *Pseudomonas*. A quantitative measurement of these compounds indicates the degree of spoilage. Fish flesh starts visibly to spoil when bacterial level rises above 10^7 organisms/ g. In chemical assessment of quality, the various products of spoilage in fish muscle are quantitatively determined and correlated with sensory characteristics. These compounds are produced in fish muscle by autolytic enzymes, putrefactive microorganisms or by chemical reactions like lipid oxidation. During spoilage, these compounds gradually accumulate in the flesh and hence their determination provides a measure of the progress of spoilage. The compounds found most useful as quality indices are:

1. Volatile bases - Basic nitrogenous compounds such as ammonia, trimethylamine (TMA), dimethylamine (DMA), etc.
2. Nucleotides - Degradation products from adenosine triphosphate (ATP) eg. inosine monophosphate (IMP), hypoxanthine (Hx), etc.
3. Lipid oxidation - Peroxides, hydroperoxides, aldehydes, etc.

Spoilage changes result in the gradual accumulation in the flesh of compounds, the quantity of which therefore provides a measure of the progress of spoilage that is independent of sensory assessment. Of the many chemical methods proposed, only those based on the measurement of amines and ammonia or on the measurement of the breakdown products of adenosine triphosphate (ATP) are of broad application in routine quality control activities.

The most well known of these compounds is the trimethylamine (TMA), volatile base, derived partly by intrinsic enzymes, but certainly by bacterial action from trimethylamine oxide (TMAO). Only marine fish species contain the TMAO and so that this test does not apply to freshwater fish and even to marine flat fish and herring, which have low concentrations of TMAO. TMA is the only one among several volatile basic compounds that increase in amount. An alternative method is to measure the total quantity of bases that can be easily volatilized, which is called as total volatile bases (TVB). The bases can be determined by making a sample of fish muscle or an extract of fish muscle alkaline, distilling off the free bases and titrating them. The bases contain one nitrogen atom per molecule and the total amount distilled is usually expressed in units of mg nitrogen / 100 g of flesh. Since TMA and TVB contain nitrogen, it is convenient to represent their concentrations in terms of mg of nitrogen in 100 g of fish flesh. Apart from distillation method, TVB can be measured on protein-free extracts by the conway method using potassium carbonate as the alkalising agent. TMA can be measured separately by a variant of the conway procedure for TVB, in which formaldehyde is added to the sample of extract before making it alkaline. The formaldehyde fixes ammonia and dimethylaime (DMA), but not TMA. A colorimetric method introduced by Dyer as Dyer's picrate procedure has been used for measuring TMA, in which formaldehyde is added to fix the DMA. The complexing of DMA is a function of the formaldehyde and hydroxyl ion concentrations and is minimized, when the extract is made alkaline with 45% potassium hydroxide rather than 50% potassium carbonate. Chemical tests based on the amounts of amines are not reliable for measuring of the freshness of the canned fish because these volatile compounds are lost to some extent in the pre-cooking stage and TMAO and other nitrogenous compounds are decomposed to methylamines and ammonia.

TVB does not increase much in the early stages of spoilage, but rises rapidly with bacterial activity. Therefore, TVB measurement is not a sensitive indicator of freshness until the fish is spoiling rapidly. It is best used to differentiate acceptable from unacceptable batches. TVB value obtained depends on the procedure used. The reason is that there are many nitrogen containing compounds in fish muscle, which decompose depending on the nature of alkali, the temperature of distillation and whether fish muscle or a protein-free extract is used, under alkaline conditions to release ammonia. TVB as measured consists of the original amines and ammonia present in the sample and the ammonia formed by decomposition during distillation. TMA content is a more precise and accurate predictor of freshness than TVB. The variance of the latter is the sum of the variances of its constituent bases and is considerably larger than that of TMA alone. TMA is also a more sensitive indicator of spoilage, since its concentration is essentially zero in fresh fish and relatively small changes can be detected with confidence. The concentration increases approximately exponentially with storage time, but can be linearised by taking its logarithm. TMA content can be used to predict storage time or freshness or quality grades. Increases in TVB and TMA during spoilage tend to be similar across a wide range of fish species, because they are produced as a result of bacterial growth, which is independent of the species. In putrid raw fish held for 20 - 25 days in melting ice, the concentrations of TMA and TVB rise to about 50 and 70 mg nitrogen / 100 g fish. For good or passable quality fish, very much lower values would be set. Not more than 1.5 mg TMAN / 100 g product has been recommended for very good quality cod for pre-packaging; 10 - 15 mg TMAN / 100 g or 35 - 40 mg TVBN / 100 g are usually regarded as the limits beyond which round, whole chilled fish can be considered too spoiled. In the case of other products, different standards operate. Thus, not greater than 30 mg TVBN / 100 g has been specified for frozen tuna and swordfish; not greater than 100 - 200 mg TVBN / 100 g for a variety of salted and dried fish; not greater than 20 mg TVBN / 100 g for the raw material used in various canned products. During storage under given conditions, changes in odour go more or less hand in hand with TMA.

An entirely different method is based on the analysis of the breakdown product, hypoxanthine (Hx) that increases in concentration in spoiling fish. Enzyme reactions continue in the tissues of fish after death, one important reaction being the degradation of adenosine triphosphate (ATP) to hypoxnthine. Hx can be measured by an enzyme method using xanthine oxidase and its concentration is used as an index of freshness. In this method, an expensive ultra violet spectrophotometer is needed. The reaction, however, can be coupled to redox indicators such as 2,6 Dichlorophenol indophenol to provide a visible colour change, which is the basis for an automated procedure, a rapid screening procedure and a simple test-strip procedure. Hypoxanthine can also be measured along with the intermediate compounds, inosine monophosphate (IMP) produced during the degradation by relatively slow ion exchange chromatography. K-value expresses the relationship between inosine and hypoxanthine and the total amount of ATP related compounds.

$$K \ (\%) = \frac{H \ x \ R + Hx}{ATP + ADP + AMP + IMP + H \ x \ R + Hx}$$

The K value of fresh fish varies from 20 to 25% and at the point of rejection; it is above 50 to 60%. It is determined by ion exchange chromatography, Reverse phase ion-pair separation by HPLC, HPLC and other methods based on the use of enzymes. Rapid high performance liquid chromatography (HPLC) procedures allow the necessary analyses to be done in less than 15 min. The advantage of analyzing all the intermediate compounds is that molar ratios of hypoxanthine or inosine plus hypoxanthine to all the ATP degradation products can be better indices of freshness than concentration of hypoxanthine. Because hypoxanthine is produced by the activity of intrinsic enzymes, its rate of increase is very variable among species. In putrid raw fish held for 20 - 25 days in melting ice, the concentrations of Hx rise to about 50 mg / 100 g fish. This value can be taken as the upper limit. In swordfish, the formation of hypoxanthine is so low that its concentration is useless as an index of spoilage. On the other hand, in ocean perch, the ATP is completely degraded to hypoxanthine within about 2 days at 0°C. However, most commercial species lie within these extremes and for these hypoxanthine can be an effective index of freshness. Maximum permissible level of Hx content in fish is 2.5 mmoles/g. In the later stages of spoilage, non-volatile biogenic amines such as histamine are formed by bacterial degradation of amino acids. Biogenic amines can be determined by chromatographic methods, and biological methods using the enzyme diamine oxidase and the substrate pyridoxal −5'-phosphate and histamine by a fluorescent method. Hypoxanthine does not seem to be affected by heat processing and so can be used as a freshness index for canned fish.

The tests, which are generally used for the chemical measurement of oxidative rancidity, are peroxide value (PV) and thiobarbituric acid (TBA) value. Oxidative rancidity, as occurs particularly in fatty fish, is a very complex deterioration in which oxygen first reacts with the unsaturated fats/lipids to form hydroperoxides, which then breakdown to substances that confer the objectionable rancid flavour. PV is a measure of the first stage determining the peroxide and TBA is the measure of the second stage determining the malonaldehyde, one of the end products of lipid oxidation. If the PV is above 10 - 20 or TBA about 1 − 2, the fish then smells and tastes rancid. The measurement of PV depends on the release of iodine from potassium iodide by the hydroperoxide and the titrimetric determination of the iodine; the units are the numbers of millilitres of 0.002 N sodium thiosulphate required to titrate the iodine liberated by 1 g fat extracted from the fish, which is equivalent to the numbers of micromoles of hydroperoxide or milliequivalents of peroxide / kg fat. TBA is measured by a rather elaborate process, the units being micromoles of malonaldehyde present in 1 g fat extracted from the sample or mg of malonaldehyde per kg of sample. Dimethylamine (DMA) can be used as a measure of the quality of frozen fish. In many species of fish, TMAO is enzymatically cleaved during frozen storage to DMA and formaldehyde (FA) and measurement of either product provides an index of cold store deterioration. DMA is best determined, along with TMA, by gas liquid chromatography (GLC). If the TMA content is low, indicating fresh fish at the time of freezing, then the presence of DMA is indicative of cold store damage. DMA is also produced during iced storage and must be allowed for; a high DMA content, if it is accompanied by a high TMA content, cannot unequivocally be interpreted as evidence of cold store deterioration. DMA can also be measured less accurately by a variation of Dyer's picrate procedure. The other product of the reaction, formaldehyde, can be conveniently measured by the Nash method using a protein-free extract of the flesh. A simple colorimeteric test has been described for cold store deterioration, which depends on the presence of formaldehyde. Methods, which depend on the production of DMA and formaldehyde, can be applied only to those species containing the TMAO splitting enzyme. This is true generally for round, white fish of commercial importance with the exception of haddock. Flat fish species do not seem to have an active enzyme. Disintegrating fish flesh in a deboning machine activates the enzyme, and high concentration of DMA can be formed in the deboned flesh before freezing. Therefore, the DMA content cannot be used as an index of cold storage changes in minced or comminuted products.

Indole production is an indication of spoilage in shrimp. Conversion of tryptophan to indole is a result of amino acid decomposition by bacteria. Indole is currently used by the USFDA to validate the sensory evaluation of shrimp decomposition. Maximum permissible level of indole in shrimps is 25 mg/100 g. Histamine formation in certain varieties of fishes such as scombroid fishes is considered as an index of spoilage. On bacterial decarboxylation of histidine, histamine is formed in fish during spoilage. Maximum permissible level of histamine in fish is 50 ppm.

The measurement of the extractability of protein from fish muscle has been used as an index of quality of frozen fish. The solubility of proteins especially salt soluble proteins gets reduced as the chilled/frozen storage period increases. The amount of soluble protein in a sample is not an absolute value as it also depends on the nature of the extracting solution and the conditions of extraction. Texture can be measured directly by instrumental methods and a simple texturometer has been used for testing the toughness of fish. Some aspects of texture, toughness, firmness and dryness are influenced by factors other than frozen storage. Cell fragility method is used for measuring toughening occurring during frozen storage. A small sample of fillet is disintegrated under standard conditions and the absorbance of the resulting suspension is used as an index of toughness. The results from this procedure too are found to be influenced by the pH of sample. Hept-cis-4-enal is considered as a direct index of the quality of a frozen product, as it contributes to the cardboardy, turnip and off-flavour during frozen storage of non-fatty fish.

An enzymatic method has been developed using malic enzyme [L- malate: NADP oxidoreductase (decarboxylating)] for distinguishing between fresh fish and frozen and thawed fish. The principle of the test is the difference in properties of two forms of malic enzyme activity in the centrifuged tissue fluid (CTF) of fish muscle. Besides the normally soluble activity (free ME), there is a latent ME activity that is readily solubilised by disruption of the tissue by a freeze-thaw cycle or by homogenization. The two forms are separable by polyacrylamide electrophoresis, and latent ME is significantly inhibited under conditions where free ME loses no activity. Free ME means malic enzyme activity present in the sarcoplasm of unfrozen whole tissue. Latent ME means the increase in malic enzyme activity effected by freezing and thawing the tissue. Two ME properties are useful for confirming this test. The first is electrophoretic mobility; the differing rates of migration of the two ME forms are demonstrated by enzymography, polyacrylamide electrophoresis coupled with a histochemical stain specific for a particular enzyme reaction. The second property is thermolability, measured by assays for ME activity before and after heat treatment of tissue fluid. Enzymography requires meticulous techniques and a full workday, and is an all or nothing test. Results from the thermolability test may be obtained in only a few hours, but are not always clear-cut, the inhibition of latent ME is not great, especially in fish aged several days on ice. If a thermolability test is positive, it is reliable as well as quick; but if the test is negative, it must be confirmed by enzymography. This enzymatic method or assay, as a measure of the rate of NADP reduction in the presence of malate. When CTF from the flesh of unfrozen haddock caught within 24 h is heated at temperatures up to 45°C, there is no loss of ME activity; but when CTF from the frozen and thawed flesh of the same fish is heated at the same temperatures, there is a drop in activity that becomes greater as the temperature increases. At higher temperatures, both free and latent ME lose activity rapidly. These tests are recommended for use only with fish that are still acceptable to the consumer, because results can be obscured by autolytic changes that set in at the beginning of spoilage.

Fresh fish can also be distinguished from frozen thawed fish by measuring the activities of enzymes like neutral b-N-acetylglucosaminidase, glutamic-oxalacetic transaminase, cytochrome oxidase, a-glucosidase and cathespin D. Besides the enzymatic methods, they are also differentiated by measuring hematocrit value of fish blood, and examining the opacity of the crystalline lens of fish eyes. The colour defect in canned tuna known as 'greening' is related to the concentration of TMAO in the flesh at the time of canning. TMAO content can be determined by measuring the TMA content in an extract before and after reduction with titanous chloride. The difference between them is the TMAO content. Tasting of canned products reveals any spoilage in the raw material before canning and any effects on flavour and texture of excess heat processing.

5.3. MICROBIOLOGICAL METHODS

Conventional and rapid methods are available for detecting the microbial quality and safety of aquatic foods.

5.3.1. Conventional Microbiological Methods

They measure the numbers of microorganisms or to indicate the presence or absence of microorganisms in a fixed quantity of products. All the recommended methods of this kind depend upon comminuting the sample very finely in a suitable aqueous medium to release the microorganisms, diluting the suspension so formed and then mixing it with a layer of agar jelly containing nutrients. When the layer in the petriplate is incubated at a suitable constant temperature between 20°C and 40°C, single individual microorganisms multiply into visible colonies that can be counted by the eye. The counts can then be related to a given unit weight of sample. Alternatively, the sample is inoculated into a special medium that on incubation indicates only whether the particular microorganism is growing. This method gives no indication of the numbers of organisms but only whether they are present or absent in a given weight of sample. Some means of swabbing the organisms to a sterile container has to be adopted before plating can be done for checking the degree of microbiological contamination of equipment, surfaces and workers. Three types of bacterial indices are used in bacteriological standards.

First is the general purpose or utility index, which expresses the total number of living bacteria present in a food. Because of the complexity and lack of uniformity in bacteriological analytical methods, this index may be variously reported as the total viable count (TVC), standard plate count (SPC), total plate count (TPC), aerobic plate count (APC) or as colony forming units (CFU). Enumeration of APC is designed to provide an estimate of the total number of aerobic organisms in a particular food. It reflects the microbiological quality of the food and is useful for indicating the potential spoilage of the perishable food products. It is also an indicator of the sanitary conditions under which the food was produced and / or processed and also of the level of Good Manufacturing Practices (GMP) adopted during processing, but in raw frozen food, uncontrolled destruction of organisms might have taken place during freezing which makes the above assumption baseless to a certain extent. Despite the limitations, APC can still be taken as a valuable indication of the effectiveness of any type of processing or chemical disinfection such as cooking, freezing and chlorination. A series of dilution of the food homogenate is mixed with an agar medium and incubated at $37^\circ \pm 1^\circ C$ for 48 h. It is assumed that each visible colony is the result of multiplication of a single cell on the surface of the agar. It is important to give incubation temperature of the bacteria, when total bacterial counts are mentioned. A bacterial count made at an incubation temperature of $20^\circ C$ is an index of spoilage or loss of eating quality, whereas a count at 35 - $37^\circ C$ is an indication of unwholesomeness or potential food poisoning. Bacterial counts are sometimes made at 25 or $30^\circ C$ in an attempt to assess both the quality and safety of food. When the TPC reaches $10^7/g$, the fish is said to be spoiled.

TVC is not a feasible measure of freshness for a number of reasons, though spoilage is a result of microbiological growth and bacterial numbers in fish increased during spoilage. The intrinsic bacterial flora on fish is predominantly psychrophilic and counts relevant to the degree of spoilage are made following incubation at $20^\circ C$ for 5 days. Only a proportion of bacteria present in a sample are active spoilers, and the rest can be considered commensal as far as spoilage is concerned; hence, TVC is only a crude measure of potential for spoilage. TVC cannot be used as a measure of the freshness of frozen stored products, because freezing and cold storage destroy a proportion of the bacterial load of the original product. Microbiological methods are fairly laborious even when necessary efforts are adopted to keep them as simple as possible. When the number of samples, which have to be examined, is large, various different kinds of automatic aids are available to facilitate the many repetitive operations. Microbiological testing is required for monitoring sanitary practices and levels of hygiene in fish processing plants and particularly in those producing cooked or heat processed products.

More specific indicators of potential public health hazards are the second group of bacteria known as organisms of public health significance or sanitary indices. They include the indicators of faecal pollution such as faecal coliforms, *E. coli*, faecal streptococci and indicator of personnel hygiene such as *Staphylococcus aureus*. Total coliforms consist of aerobic and facultative anaerobic Gram-negative, nonsporing rods that ferment lactose in Brilliant green lactose bile broth or MacConkey broth within 48 h at 37± 1°C. The predominant aerobic bacterial flora of the large intestine of humans and animals is composed of nonsporing, non-acid fast, Gram-negative bacteria. They exhibit general morphological and biochemical similarities and are grouped together in the large and complex family of Enterobacteriaceae. Members of the coliforms, including faecal coliforms are referred to as indicator organisms, since their presence in certain numbers, may indicate the potential presence of pathogens in foods. An elevated temperature is used to differentiate coliforms of faecal and non-faecal origin. Faecal coliforms ferment lactose within 48 h at 45.5 ± 0.5°C, whereas non-faecal coliforms cannot do so. Faecal coliforms normally grow in the gastero-intestinal tract of humans and other warm-blooded animals. They include *Escherichia, Klebsiella* and *Enterobacter*. This group comprises the faecal coliforms such as *Escherichia coli* and faecal Streptococci or Enterococci. These bacteria are normally present in large numbers in humans and other warm blooded animal faces and their presence in fishery products is an implication that the product has been unhygienically handled.

Faecal streptococci or enterococci are gram-positive non-spore forming and non-motile cocci, which are found in humans and animal faeces in large numbers, and hence their presence in food has been well accepted as an indicator of faecal contamination. In addition to their faecal sources, these organisms exist in plants, insects and soils. Just like *E. coli*, faecal streptococci are also absent in offshore waters. Unclean boat deck, utensils, water and ice are the major sources of streptococci contamination to the product. Faecal streptococci are comparatively resistant to many adverse conditions. About 30% reduction of faecal streptococci takes place during freezing (-40°C) whereas during subsequent storage at –20°C, not much of reduction takes place in count. Moreover, faecal streptococci are also useful in determining the post-process proliferation of faecal contamination in foods, which cannot be detected by the use of much less resistant *E. coli*. A large number of investigators have examined the use of an Enterococcus index for food safety and feel that it is better index of food sanitary quality than the coliform index, especially for frozen foods. Faecal streptococci count in frozen shrimp should not exceed 100 cfu/g. Many methods have been proposed for the isolation and enumeration of the streptococci and most procedures employ presumptive media followed by confirmatory tests. KF agar medium is a selective, differential medium that contains sodium azide as the major selective agent and triphenyl tetrazolium chloride (TTC) for differential purposes. The medium also contains a relatively high concentration of maltose (2.0%) and a small amount of lactose (0.1%). *Streptococcus faecalis* and its varieties produce a deep red colour when TTC is reduced to its formazan derivative. Other species are not as reductive, thereby producing light pink colonies.

Staphylococci are gram-positive cocci belonging to the family Micrococcaceae. *Staphylococcus* is expected to exist, at least in low numbers in any or all food products that are of animal origin; or in those that are handled by humans. It has been estimated that about 30% of humans are carriers of *S. aureus*. Palms of the workers having some wounds or cuts in unprotected condition may harbour many thousands of *S. aureus*, which will contaminate the material during handling. Unnecessary talking during processing may result in the expellation of saliva containing *S. aureus*. This may fall on the food material resulting in contamination. During freezing at –40°C, 5-10% of the organism is destroyed and then during frozen storage it gradually disappears. The most obvious control of *S. aureus* in food is to raise the hygiene of the workers and to enforce adequate control over the holding conditions like time and temperature. The maximum allowable limit of coagulate-positive *S. aureus* in frozen shrimp is 100 cfu/g. Direct plating for the enumeration of *S. aureus* is suitable for the analysis of foods in which more than 100 *S. aureus* cells/g are expected. Food samples may be subjected to a variety of media in various ways in applying the direct plating technique. A selective chemical more frequently used in media for the isolation and enumeration of staphylococci is potassium tellurite (K_2TeO_3). Of the media formulated for the isolation and recovery of staphylococci, Baird Parker agar (BPA) appears to be the most widely used as a direct plating medium as it is more selective, does not inhibit the growth of injured cells and it clearly supports the growth of *S. aureus*. *S. aureus* colonies on Baird Parker agar are usually 1.5 mm in size, jet black to dark grey, smooth, convex, have entire margins and may show an opaque zone and/ or clear halo extending beyond the opaque zone.

The third group consists of the specific food poisoning bacteria such as *Salmonella, Listeria monocytogenes, Clostridium botulinum, Vibrio cholerae* and *V. parahaemolyticus*. Of these, only *Salmonella* and *V. cholerae* are commonly incorporated in standards for fishery products. The International Commission on Microbiological Specification for Foods (ICMSF), has published protocols for the detection and enumeration of microorganisms in food including fish and fishery products, which have the merit of some degree of international recognition. Routine testing for sterility of canned products is almost always unnecessary; and the process itself should be carried out so as to ensure an adequately low incidence of unsafe or spoiled containers. Adequate sterility results in complete softening of the bones of even large fish. Therefore, bone softening offers a simple guide to the effectiveness of processing. If sensory inspection of contents reveals any hardness, that batch of products should be examined for sterility. Anaerobic sulphite reducers are another group of microorganisms occasionally incorporated in the microbial standards for fish and fishery products. The maximum permissible limit for this organisms in frozen squid and cuttlefish is 100 /g. It is enumerated by MPN technique using differential reinforced clostridial medium (DRCM) with an overlay of liquid paraffin.

5.3.2. Rapid Microbiological Methods

Few rapid methods of assessing microbial quality of fish are dye reduction test, limulus amoebocyte lysate test and electrical impedence test.

i. Dye reduction test

This test has been in use in assessing the bacteriological quality of milk for a long time. Resazurin and methylene blue are the most commonly used dyes. It has been observed that approximately 10^4 bacteria / ml reaction mixture is required to cause reduction of resazurin in 8 h. Methylene blue reduction test has been applied for approximation of bacterial counts in shrimps and oysters. Resazurin reduction test has been applied for microbiological control of deep frozen shrimps and found good correlation between reduction times and viable counts. A reduction time of 4 h discriminated well between samples that complied or not, employing with the normally used bacteriological limit of 10^5 cfu / ml. The dye reduction test rejected three of the fifty samples, which had acceptable plate counts, but no false negatives were recorded.

ii. Limulus amoebocyte lysate (LAL) test

Mostly gram-negative bacteria (GNB) are responsible for the spoilage of fish when they are in raw / fresh condition. The GNB except *Shigella dysenteriae* and *Vibrio cholerae*, are characterized by their production of endotoxins, which consist of lipopolysaccharide (LPS) layer of the cell envelop. The gram-negative flora of fish can be determined by plating procedures employing bile salts containing media, and this method depends on the cfu capacity of the cells under the cultural and incubation conditions employed. A 24 h incubation period is necessary for results, with longer times being required for psychrotrophic types. The use of a highly sensitive and rapid test known as Limulus Amoebocyte Lysate (LAL) test to measure the endotoxins of gram-negative bacteria in foods can be carried out in a shorter period of time with only an hour incubation required. LAL is an aqueous extract of blood cells called amoebocytes from the horseshoe crab, *Limulus polyphemus*. LAL employs a lysate protein obtained from the blood or haemolymph cells of the horseshoe crab. This lysate protein is most sensitive to endotoxins. The lysate test for endotoxins consists of adding aliquots of food suspensions (0.1ml) or other test materials to small quantities of a limulus lysate preparation (0.1 ml) followed by incubation at 37°C for 1 h. The presence of endotoxins causes gel formation of the lysate material. Since both viable and non-viable gram- negative bacteria are detected by the LAL test, simultaneous plating is necessary in order to determine the number of viable organisms. But, it has been found that endotoxin titres increased in proportion to the viable counts of gram-negative bacteria.

iii. Electrical impedence test

Impedence is the total electrical resistance to the flow of an alternating current through a given medium, conducting material, and has been shown to be a complex entity composed of a resistive or conductive component and a reactive component, capacitance. Impedence technique relies on the fact that metabolizing microorganisms alter the chemical composition of the growth medium and that these chemical changes cause a change in the impedence of the medium. Impedence and conductance measurements determine the microbial content of a sample by monitoring microbial metabolism rather than biomass. As microorganisms grow, they utilize nutrients in the medium converting them into smaller highly charged molecules such as fatty acids, amino acids and various organic acids. When impedence of broth cultures is measured, the curves that result resemble bacterial growth curves. The signals or curves are reproducible for species and strains and mixed cultures can be identified by the use of specific growth inhibitors. Impedence changes are detectable when the concentration of microorganisms exceeds a threshold level of 10^6 - 10^7 cells / ml. The time required for the initial inoculum to reach threshhold level is designated as the detection time and is a function of both the initial concentration and specific growth kinetics of the organism in a given medium. By comparing the detection time obtained to the results of a standard calibration curve, an estimate of the initial concentration of the microorganisms can be made. This technique has been automated and 'Bactometer' is an instrument developed based on this principle.

5.4. STATISTICAL METHODS

Food quality cannot be properly measured unless the correct number, size and kind of sample are selected. For several reasons, it is not possible to examine every item in a batch of product and a decision as to whether to accept or reject a batch, whether to place a batch in a particular grade, must be based on a sample taken from the lot presented for inspection and hence, the statistical principles of sampling and of decision making based on samples must be known by the quality control personnel. Sampling inevitably entails a risk that the result of the measurement does not represent the true nature of the batch. The larger the number of samples, the lower the risk. The number of samples taken for examination from a batch may have to be moderated by the costs involved. Costs arise from the actual labour of sampling and of examination, but if the method used is destructive then also from the lost product. End product inspection of fish products uses the practices of acceptance sampling by attributes. A sample of 'n' items is taken from the lot and tested according to the product specification. After examination, each item is classed as having passed or failed. The lot is accepted if 'c' or less items in the sample fail. The sampling plan specifies values of 'n' and 'c'.

The Codex Alimentarius has recommended sampling for use in the inspection of pre-packaged foods including fish products. The sampling plans are based on Acceptable Quality Level (AQL) of 6.5%. Inspection of fish products almost invariably means loss of the samples from the batch. Even if they were not physically destroyed during analysis, they would not normally be in a fit state to be returned. Thus, the value of the samples must be added to the direct costs of examination. The lowest level of sampling in the Codex sampling plans, n = 6, c = 1, does not give much protection against rejecting a low quality batch. The probability under this plan of accepting a lot with 30% defectives is 0.42. This probability is halved, though at the expense of approximately doubling the cost, with a sampling plan of n = 13 and c = 2. Attributes sampling plans are appropriate for inspecting fish products, whose quality is defined in standards like those of the Codex Alimentarius in which a demerits points system is used. The sample unit is inspected for defects and the defects are allocated points according to the nature of the defect.

A unit is failed, if the total demerit points exceed the control value. In grading schemes, grade boundaries are defined by the number of demerit points. Fish can also be inspected and graded using variables sampling plans. These are appropriate to situations, where the quality factor of interest can be measured on a scale, for example, weight or freshness and the grade boundaries or limits of acceptability can be defined by values on the scale. Similar statistical principles of sampling and decision making apply to inspection of fish for microbiological quality. Appropriate three-class sampling plans for a wide range of food products, including fish, have been given by the International Commission on Microbiological Specification for Foods (ICMSF). Statistical Quality Control (SQC) is a very useful technique for large scale production. The control charts or Shewhan charts used in SQC show where and when the production process has gone out of control and indicate the lines on which action should be taken. Sampling methods are random sampling, approximation to random selection, stratified sampling and sampling in stages. Difference between sampling inspection and 100% inspection is given below:

	Sampling Inspection	100% Inspection
1.	Less expensive	More expensive
2.	If the test is destructive, it is preferable	Possible only for non-destructive tests
3.	When some risk can be taken	When no risk can be taken
4.	More detailed examination is possible	Less accuracy, less detailed examination
5.	Less error	Inspection fatigue errors

CHAPTER 6

HACCP FOR AQUATIC FOOD INDUSTRIES

Hazard analysis critical control point (HACCP) system identifies, evaluates and control hazards, which are significant for food safety. HACCP system, thus, clearly identifies food safety problems and also where and how these problems can be controlled or prevented. Preventative or control measures have to be described and persons, who are responsible for their execution, have to be trained in order to assure that these actions are executed regularly and consistently. HACCP was originally developed and used by the private food industry, Pillsbury Company in the late 1960's for the safety of the food intended for the space programme conducted by the US National Aeronautics Space Administration (NASA). The HACCP concept was publicly presented at the US National Conference on Food Protection during 1971. A book on HACCP was published in 1988 by the International Commission on Microbiological Specifications for Foods (ICMSF). The US National Advisory Committee on Microbiological Criteria for Foods (NACMCF) approved the first major document on HACCP in 1989. NACMCF issued a revised document on HACCP in 1992 by incorporating seven principles into HACCP system. Codex issued the first HACCP Guidelines in 1993 that were adopted by the FAO/WHO Codex Alimentarius Commission. Based on a number of FAO/WHO Consultations, Codex issued a revised document in 1997. Simultaneously NACMCF issued the third revised document, which is similar to revised document of Codex. HACCP was finally integrated into the official regulations of the European Union (EU) in 1994 through its Commission Decision 94/356/EEC detailing the rules for the application of the HACCP system and the United States Food and Drug Administration (USFDA) in 1995 by issuing the Code of Federal Regulations on safe and sanitary processing and importing of fish and fishery products.

There are some differences in the HACCP systems practiced in EU and US, despite the fact that both the systems have seven principles. These differences are mainly related to the prerequisites programmes, the way they are documented and verified, and the scope and content of the identification of hazards. According to the Agreements of the World Trade Organization (WTO) viz. Agreement on Sanitary and Phytosanitary (SPS) Measures and the Agreement on Technical Barrier to Trade (TBT) made in April 1995, the work of Codex is recognized as the reference for international food safety requirement. This implies that all the member States of WTO cannot reject food, which meets Codex recommendations and standards without providing justification based on risk assessment. Since the application of HACCP is recommended by Codex, the HACCP has become the international reference system for food safety assurance. Seven basic principles of HACCP system are

1. Conduct a hazard analysis
2. Determine the critical control points (CCPs)
3. Establish critical limits
4. Establish monitoring procedures
5. Establish corrective actions
6. Establish verification procedures
7. Establish record-keeping and documentation procedures

It is necessary to know that the food sector should be operating on the basis of a prerequisite programme, prior to the application of HACCP to that food sector. It is also essential that top management is fully committed to introduce the system. Many departments and different personnel from Chief to ordinary workers should be involved and responsible for part of the system, and their full support and cooperation are needed. The Codex Guidelines suggest that the introduction and application of the HACCP principles should follow a series of 12 steps in a logical sequence, which is given below:

Step 1: Assemble the HACCP team

Introduction of a HACCP system in large food industries is a complex process and requires a multidisciplinary approach by a team of specialists. The microbiologist is of paramount importance, and must advise the team on all matters related to microbiology, safety and risks. He/she must have an updated knowledge on these matters and also access to technical literature on the most recent developments in his/she field. In many cases, he/she will also need access to the use of a well-equipped laboratory if specific questions and problems cannot be solved by studying the technical literature. Examples are investigations on the microbial ecology of specific products, challenge tests and inoculation studies for evaluation of safety aspects. Another important member of the HACCP team is the processing specialist. He/she must advise on production procedures and constraints, prepare the initial process-flow diagram, advise on technological objectives at various points in the process and on technical limitations of equipment. Other technical specialists such as a food chemist, a food engineer as well as packaging technologists, sales staff, training and personnel managers can provide valuable information to the HACCP team and they should attend some of the meetings. Key members of the HACCP team (including the leader) must have an intimate knowledge of the HACCP system. Small and medium size industries are not likely to have qualified personnel on the payroll and must therefore get assistance from outside consultants in order to implement the system. One person should be appointed as leader of the team. When the HACCP team is assembled, the scope of the HACCP plan should be identified; describing which segment of the food claim is involved and addressed in the work.

Step 2: Describe product

A full and detailed description of the final production must be drawn up. The raw materials and ingredients used must be specified including the market name or Latin name of the fishery component. Details regarding hazards in the raw material will be included in the HACCP plan. All factors which influence safety such as composition, physical/chemical structure including water activity (a_w) and pH must be described, and any bactericidal treatments such as heating, freezing, brining and smoking must be specified as well as packaging type, storage conditions and methods of distribution. The normal shelf life under specified condition should also be recorded.

Step 3: Identify intended use and consumer

The HACCP team needs to identify the intended use and consumer of the product. The consumer should specify the intended use or expected use. The use and preparation greatly influence the safety of the product. Certain products may be contaminated or carry pathogenic organisms as a part of the natural flora. The intended consumer may be the general public or a particular segment of the population such as infants or elderly. If the product is to be sold to hospitals or groups of the population with high susceptibility, more safety is required and critical limits need to be stricter.

Step 4: Construct flow diagram

The purpose of the flow diagram is to provide a clear simple description of all steps involved in the processing. Receiving and storage steps for raw materials and ingredients should be included. Time and temperature conditions during processing should be mentioned whenever there is a holding step e.g. in holding vats, buffer tanks or other areas.

Step 5: On-site confirmation of flow diagram

The constructed flow diagram should be verified on-site for accuracy. The site should be inspected during all hours (night shifts, weekends) of operation to check for correctness and ensure that nothing crucial was overlooked.

Step 6: List all potential hazards associated with each step in the operation, conduct a hazard analysis and consider any measure to control identified hazards (Principle 1)

Hazard refers to both a specific agent and/or a condition with the potential to cause harm. After having identified all potential hazards all the information available must be evaluated in order to decide, which of the hazards are significant and reasonably likely to cause illness if not effectively controlled. The hazard analysis is the key to preparing an effective HACCP plan and serves the following purposes:

1. Hazards and associated control measures are identified
2. Needed modifications to a process or product is identified
3. Provides a basis for determining CCPs (Principle 2)

The following questions to be considered when conducting a hazard analysis:

1. Do the raw materials and ingredients contain any hazardous agents?
2. Does the food permit survival, multiplication of pathogens or toxin formation?
3. Are any pathogens destroyed during processing, and are there any possibilities for recontamination?
4. Does the packaging affect the microbial population?
5. Does the consumer heat the food?
6. Is the product for the general public or for consumption by a population with high susceptibility to illness?

A decision tree with a number of questions can be used to determine, if potential hazards are 'real'. The questions have to be asked at each step of the processing chain and all hazards must be considered. An element of risk assessment is involved in the evaluation of potential hazards. Only these hazards, which are likely to occur, and which will cause a reasonably serious adverse health effect are regarded as significant. Thus, the basic procedures to use in conducting the hazard analysis is as follows:

- Based on the product description and the flow diagram, all the potential hazards associated with the product and at each processing step is determined and listed
- Make a hazard evaluation:
  o assess severity of health consequences if potential hazards are not controlled
  o determine likelihood of occurrence of potential hazards if not properly controlled

o using information above, determine if this potential hazard is to be addressed in the HACCP plan

o describe control measures

Control measure is any factor or activity, which can be used to prevent, eliminate or reduce a food safety hazard to an acceptable level. More than one control measure may be required to control a hazard. Upon completion of the hazard analysis, the hazards associated with each step in the production should be listed along with any measure that is used to control the hazards. A "hazard analysis worksheet" can be used to organize and document the considerations in identifying food safety hazard.

Step 7: Determine the critical control points (CCPs) (Principle 2)

Complete and accurate identification of all the CCPs is fundamental to controlling food safety hazards. To facilitate this identification, the use of a CCP decision tree can be of great help. The first two questions deal with the raw material. It is important to note, that if an identified hazard is eliminated or reduced at a later process step or by normal consumer use, the raw material is not a CCP. Third question deals with formulation or composition of the product in preventing multiplication of pathogens. Fourth question asks if contamination, recontamination or even multiplication of pathogens can take place at this step. If the answer is 'No', sixth question "Is the process step intended to eliminate or reduce the hazard to an unacceptable level?" thus has to be answered, but if the answer is 'Yes', the answer to fifth question "Will subsequent processing steps, including expected consumer use, guarantee removal of the hazard or reduction to an acceptable level?", will decide whether this step is a CCP or not. Only points where truly significant hazards can be controlled should be designated as CCPs. A tendency exists to control too much and to designate too many CCPs. This should be avoided as it will create confusion and divert attention from the true CCP.

Step 8: Establish critical limits (Principle 3)

All critical limits should be scientifically based and refer to factors such as: time/temperature conditions, moisture level, water activity

(a_w), pH, titratable acidity, salt concentration, available chlorine, preservatives, sensory quality. Microbiological limits should normally be avoided. This is because microbiological data can usually only be produced by a process, which may take several days. The monitoring of microbiological limits would therefore not allow you to take instant action when the process deviates. Authoritative critical limit information is available from sources such as the "Fish and Fisheries Products Hazards and Control Guide" of USFDA or obtained from regulatory agencies. When critical limits have been established, they should be entered on the HACCP Plan Form.

Examples of critical limits

Hazard	CCP	Critical limit
Bacterial pathogens (Non-sporulating)	Pasteurization	72°C for at least 15 sec.
Metal fragments	Metal detector	Metal fragments larger than 0.5mm
Bacterial pathogens	Drying oven	a_w < 0.85 for controlling growth in dried food products
Excessive nitrite	Curing room/brining	Max. 200 ppm sodium nitrite in finished product
Bacterial pathogens	Acidification step	Max. pH of 4.6 to control *C. botulinum* in acidified food
Food allergens	Labelling	Label that is legible and contains a listing of correct ingredients
Histamine	Receiving	Max. of 25 ppm histamine levels in evaluation of tuna for histamine

Critical limits Vs Operating limits

Process	Critical limit	Operating limit
Acidification	pH 4.6	pH 4.3
Drying	a_w 0.84	a_w 0.80
Hot fill	80°C	85°C
Slicing	2.0 cm	2.5 cm

Step 9: Establish monitoring procedures (Principle 4)

The main purpose of monitoring is to determine if there is loss of control or deviation. Monitoring of CCPs serves the following purposes:

- to determine if there is a loss of control and a deviation occurs at a CCP. Appropriate action must then be taken
- monitoring keeps check on the operation and provides information whether there is a trend towards loss of control and action can be taken to bring the process back into control before a deviation occur
- provides written documentation for use in verification and audit. All records must be signed.

All monitoring must be done rapidly and results must be evaluated by a designated person with knowledge and authority to carry out corrective actions. Monitoring methods typically include time/temperature recording, pH and a_w measurements and sensory quality.

Step 10: Establish corrective actions (Principle 5)

Whenever there is a deviation from established critical limits, a corrective action must be instituted to ensure that defective products do not reach the consumer. These actions should include the following:

- determine and correct the cause of deviation
- determine the disposition of products that were produced during the process deviation
- record the corrective action taken

Options for disposition of products placed on-hold are isolating and holding products for safety evaluation, reprocessing, rejecting and/or destroying of product and use as by-product. Corrective action procedures should be developed by the HACCP team in advance and specified in the HACCP plan. Any action should be recorded on the HACCP Plan Form. If necessary, a more detailed corrective action report should be elaborated including the information on product identification, description of the deviation, results of the product evaluation, corrective action taken including the final disposition of the affected product, actions to prevent the deviation from recurring and name of the individual responsible for taking action.

Step 11: Establish verification procedures (Principle 6)

The purpose of the HACCP plan is to prevent food safety hazards from occurring. Verification activities must provide a level of confidence that the HACCP plan is working properly and is adequate to control hazards. The following elements should be included in the verification activities:

- Validation – initial and subsequent validation of the HACCP plan
- Verification of the CCP-monitoring
  - o CCP-record review
  - o Calibration of instruments
  - o Targeted sampling and testing
  - o Microbiological testing
- Review of monitoring, corrective action records
- Comprehensive HACCP system verification

Thus, the verification procedures include verification of both the individual CCP and the overall HACCP plan. An essential component of verification is validation. In validation of the HACCP plan, it needs to be established that the plan is scientifically and technically sound. This means that scientific validation includes review of each part of the HACCP plan from the hazard analysis through to each CCP. The needed information can be obtained from expert advice, scientific studies and literature, in-plant observations and measurements. Apart from the initial validation, subsequent validation as well as verification must take place whenever there is a change in raw materials, product formulation, processing procedures, consumer and handling practices, new information on hazards and their control, consumer complaints, recurring deviations or any other indication, that the system is not working. A periodic comprehensive verification of the HACCP system should be conducted yearly by an unbiased, independent authority. This should include a review of the HACCP plan for completeness, confirmation of the flow diagram, review of all records and validations, sampling and testing to verify CCPs. Verification is the responsibility of the producer or food handler. However, where regulatory agencies are conducting audits or sampling end products the results can be used by industry as part of the verification programme. Verification procedures should be entered on the HACCP Plan Form and results into special verification records.

Step 12: Establish record-keeping and documentation procedures (Principle 7)

Records and documentation are vital for the verification and auditing to determine, if the HACCP system in operation is in compliance with the HACCP plan. Records of support documents must also be kept such as data used to establish critical limits, reports from consultants or experts, a list of the HACCP team and their responsibilities and the preliminary steps taken before development and implementation of the HACCP plan. Documentation includes hazard analysis worksheet, CCP determination and critical limit determination. Examples of records are CCP monitoring activities, deviations and associated corrective actions and modifications of the HACCP system.

HACCP Audit

Many fish producing, exporting and/or importing countries have undertaken a thorough evaluation and reorganization of fish inspection and quality control systems with the aim to improve efficiency, rationalize human resources and introduce risk analysis-based approaches. The HACCP principles play a pivotal role in these preventive approaches. Their application is the responsibility of the fish industry, whereas government control agencies are responsible for monitoring and assessing their proper implementation. Many inspection agencies have developed approaches and procedures for carrying out HACCP compliance auditing. These approaches and modalities have used the terminology and basic requirements of the ISO 10011 standards that were adapted to the specifications of HACCP and to the countries regulations.

Audit is a systematic and independent examination to determine whether activities and results comply with the documented procedures; also whether these procedures are implemented effectively and are suitable to achieve the objectives. In HACCP terms, achieving the objectives means managing the production and distribution of safe fish products through the use of an HACCP based approach. The outcome of the audit is to establish whether the manufacturer has implemented a sound HACCP system, the knowledge and experience needed to maintain it and the necessary support (or prerequisite) programmes in place to assess adherence to Good Hygienic and Good Manufacturing Practices (GHP/GMP). The audit encompasses assessment of the management commitment to support the system and assessment of the knowledge, competency and decision-making capabilities of the HACCP team members to apply the system and maintain it. Four types of HACCP audits can be envisaged:

- An internal HACCP audit to establish the effectiveness of the HACCP system using the company's own human resources or by bringing in an external HACCP assessor.

- An external HACCP audit of suppliers of critical raw materials or of packed finished products to establish whether they have robust HACCP system in place. This includes regulatory HACCP auditing.

- Audit of the customers HACCP system. This may be important where the customer is responsible for the distribution and sale of a high risk (e.g. a chilled ready meal) product which bears the brand of the manufacturing company

- An investigative audit can also be conducted to analyze a specific problem area. This may be used for example when a CCP regularly goes out of control and more studies are needed to investigate the real cause in order to take corrective action.

Application and implementation of HACCP in the fish industry

It is generally accepted that responsibility for producing safe food is in the hands of the producer. It is therefore the responsibility of the producer to ensure the development and application of a proper HACCP plan. It is of great importance that senior management of a company needs to understand and support the implementation of HACCP in the processing facilities. They need to understand the benefits as well as the cost and the resources needed. While the HACCP team is conducting the HACCP study, it is advisable to initiate training of key personnel. No plan will work if the people who have to implement it are not trained. People who have to develop the plan as well as people responsible for implementation and maintenance may all need training. When the HACCP plan has been developed, it needs to be approved by senior management. During the development of the HACCP plan including the prerequisite programme, it often becomes clear that improvements may be needed in construction or layout of facilities, or utensils is need to be replaced. This could involve considerable costs and run into budgetary constraints. However, modifications, which are essential to food safety, should always be executed immediately, while a timetable for less necessary modifications should be made.

Although implementation of HACCP is the responsibility of the industry, government authorities also have a role to play. Government authorities can play three roles viz. as facilitators, enforcers or trainers.

· As facilitators, they can help industries understand the goals and scope of HACCP and provide expertise during the establishment of a HACCP plan or its verification

· As enforcers, their task is to assess the correct application and implementation of the seven HACCP principles

· As trainers, they can provide training courses and also participate in training courses organized by or for the industry

Hence, it is a key role of government agencies to show leadership by promoting and facilitating the implementation of HACCP. However, the government has also a strategic role as well as an ongoing role in assessing the HACCP systems applied in the industry. With regard to the actual assessment of HACCP, government agencies also play an important role in providing guidance on the assessment process needed to be developed and provided to officials for its uniform and acceptable application. This guidance should be developed by government agencies in collaboration with, when possible, food control officials and industry. Proper implementation of HACCP may also need the support of other institutions such as academia and research, trade associations, private sector, etc. A consequence of the WTO/SPS agreement is that food safety criteria such as FSOs, performance criteria and microbiological end product criteria have to be based on scientific evidence, and where appropriate on a risk assessment. Finally, consumers and consumer advocate groups have a counter-balancing role to ensure that safety and quality are not undermined by political and socio-economical considerations when drafting legislation or implementing safety and quality policies.

The safety of seafood products varies considerably and is influenced by a number of factors such as origin of the fish, microbiological ecology of the product, handling and processing practices and preparations before consumption. Taking most of these aspects into consideration, seafood can be conveniently grouped into nine categories. They are:

1. Molluscan shellfish
2. Raw fish to be consumed without any cooking
3. Fresh or frozen fish and crustaceans – to be fully cooked before consumption
4. Lightly preserved fish products i.e. NaCl <6% in water phase, pH >5.0. The prescribed storage temperature is <5°C. This group includes salted, marinated, cold smoked and gravid fish
5. Fermented fish, i.e. NaCl <8% NaCl, pH changing from neutral to acid. The products are typically stored at ambient temperature
6. Semi-preserved fish i.e. NaCl >6% in water phase, or pH <5, preservatives (sorbate, benzoate, nitrite) may be added. The prescribed storage temperature is <10°C. This group includes salted and/or marinated fish or caviar, fermented fish (after completion of fermentation)
7. Mildly heat-processed (pasteurized, cooked, hot smoked) fish products and crustaceans (including pre-cooked, breaded fillets). The prescribed storage temperature is <5°C
8. Heat-processed (sterilized, packed in sealed containers)
9. Dried, smoke-dried fish, heavily salted fish. Can be stored at ambient temperatures

Hazard characteristics and risk factors

The following six hazard characteristics and risk factors have been considered in ranking aquatic food into risk categories:

1. No terminal heat treatment

Apart from raw fish to be consumed cooked or fried, all other fish products are ready-to-eat.

2. Safety record

The safety record is poor for molluscan shellfish and fish to be consumed raw due to the presence of accumulated biological hazards (viruses, pathogenic bacteria, parasites, biotoxins), molluscan shellfish, tropical reef fish and scombroid fish to be cooked before consumption due to the presence of heat stable aquatic toxins or scombrotoxin, presence of heat stable biogenic amines in canned sterilized products and few outbreaks of botulism caused by the same type of product and some fermented fish; e.g. salted fish from the Middle East or products from Alaska.

3. No critical control point for at least one identified hazard in production/processing

The situation applies to the accumulation of biological hazards in shellfish and presence of biotoxins (ciguatera) in fish from tropical reef.

4. Product is subject to potentially harmful conta-mination or recontamination after processing and before packaging

All raw fish and fishery product, which has not been subject to any bactericidal treatment, are likely to harbour pathogenic organisms as part of their natural flora. Potentially harmful recontamination is possible and reasonably likely to occur for products being mildly heat-treated before being placed in the final container (cooked shrimp, hot smoked fish). However, the risk associated with lightly preserved fish, and fish and shellfish to be consumed raw may also increase due to this factor (e.g. contamination of cold smoked fish with *L. monocytogenes*).

5. Products with a potential for abusive handling

This hazard refers mainly to handling and storing the fish product at abuse (elevated) temperatures. With the exception of sterilized, canned or fully preserved products, there is a potential for this hazard for all other types of fish products. However, this is not likely to occur or fish to be consumed raw, as spoilage will be very fast at elevated temperatures.

6. Growth of pathogens

The growth of pathogens, particularly in ready-to-eat products is a serious hazard. Two potential hazards of this nature are known and likely to occur: the possible growth of *L. monocytogenes* in lightly preserved fish products and the growth of *C. botulinum* in some types of fermented aquatic foods. Growth of other pathogens in preserved or heat-processed products is possible only if the preserving parameters are not applied as specified and other potential hazards are in fact occurring (temperature abuse, recontamination of heat processed fish). Spoilage bacteria grow in all types of fish products (except sterilized products) and in most cases they grow faster than any pathogen. This is particularly the case in raw, unprocessed or unpreserved fish. The growth of pathogens is not considered an additional hazard likely to occur and influence the safety of this product.

Different aquatic food are assigned to a risk category in terms of health hazards by using a "+" (plus) to indicate a potential risk related to the hazard characteristics. The number of plusses then determines the risk category of the seafood concerned.

CHAPTER 7

PLANT SANITATION AND HYGIENE

Any food safety programme shall include both the HACCP and sanitation control procedures (SCP) for its successful implementation. SSOPs industry SCPs, GHPs, and GMPs are prerequisites for the implementation of HACCP system. Similar to documenting procedures in a HACCP plan, sanitation standard operating procedures (SSOP) outline how an industry maintains sanitary control within the plant. Although a written SSOP plan is not required by the US FDA, it is recommended to explain the in-plant procedures the industry will follow to control, monitor and correct the key sanitation conditions and practices.

7.1. SANITATION STANDARD OPERATING PROCEDURES (SSOP)

SSOP plans are recommended because they describe the sanitary procedures to be used in the plant; provide the schedule for the sanitation procedures; provide a training tool for employees; identify trends and prevent re-occurring problems; ensure that everyone, from management to production workers, understands acceptable sanitation practices; provide the foundation to support a routine monitoring programme; encourage prior planning to ensure that corrections are taken when necessary; demonstrate commitment to consumers and inspectors; and lead to improved sanitary practices and conditions in the plant.

Like HACCP plans, SSOP should be specific to each plant. SSOP should describe the plant's procedures associated with sanitary handling of food and the cleanliness of the plant environment and the activities conducted to meet them. Plants can choose to develop informal or formal SSOP plans. Informal SSOP may simply outline the frequency and procedures to be followed to control, monitor and correct deficiencies for a specific task or sanitation concern. Formal SSOP are written to follow a standard format, so each SSOP is developed to contain standard information. Prior to developing a formal SSOP plan, industries would design a standard format to use for each individual SSOP. The standard format may include some or all of the following sections:

1. Purpose or objective of the SSOP

2. Scope or relevance of the SSOP (eg. preparation of hand dip stations in product packing room)

3. Responsibility (eg. the individual or job description responsible for implementing and / or monitoring the procedures in the SSOP)

4. Materials and equipment (eg. listing any special tools or equipment needed to carry out the task and / or monitoring activity)

5. Procedures (documentation of the procedures necessary to carry out the SSOP) and frequencies (how often the procedure in the SSOP will be used)

5. Documentation of changes (records why changes were made to SSOP and documents version numbers and so the most recent version is being used)

6. Approval section (eg. signatures of acceptance by plant management)

The two most important aspects of any type of SSOP – either informal or formal – is that enough details is provided for someone to carry out the task in question, and the procedures listed accurately reflect the activities that are being conducted. An SSOP with too much detail may be counter-productive because strict adherence to the procedures may be difficult to achieve every time and it is likely to be informally modified over time. Likewise, an SSOP without enough detailed information will not be useful for a plant.

An easy way to start writing SSOP is to think through each sanitation operation that is being conducted in the plant and document how it is conducted, where it is being conducted, and who is responsible for conducting the operation. In addition, think through how the sanitation control procedure will be monitored, recorded and corrected if there is a deviation. Simply writing down the sanitation procedures that are currently being conducted in the plant is the first step to developing an SSOP plan.

The SSOP sections are based on the eight FDA key sanitation conditions. Although the approaches may differ, sanitation procedures, monitoring requirements, and necessary corrections all emphasize the importance of a written SSOP plan to support and explain the monitoring requirements and necessary corrections. SSOP plans will vary from plant to plant because each plant and process is designed differently.

7.2. SANITATION CONTROL PROCEDURES (SCP)

It is a procedure to maintain sanitary conditions usually related to the entire processing facility or an area, not just limited to a specific processing step or CCP. Each processor should have and implement a written sanitation standard operating procedure or similar document that is specific to each location where fish and fishery products are produced. The SSOP should specify how the processor would meet those sanitation conditions and practices that are to be monitored.

Each processor shall monitor the conditions and practices during processing with sufficient frequency to ensure, at a minimum, conformance with those conditions and practices specified in Good Manufacturing Practices (GMP) that are both appropriate to the plant and the food being processed and relate to the following FDA key sanitation conditions.

1. Safety of the water that comes into contact with food or food contact surfaces, or is used in the manufacture of ice

2. Condition and cleanliness of food contact surfaces, including utensils, gloves and outer garments

3. Prevention of cross-contamination from unsanitary objects to food, food packaging material, and other food contact surfaces,

including utensils, gloves and outer garments, and from raw product to cooked product

4. Maintenance of hand washing, hand sanitizing, and toilet facilities

5. Protection of food, food packaging material, and food contact surfaces from adulteration with lubricants, fuel, pesticides, cleaning compounds, sanitizing agents, condensate and other chemical, physical and biological contaminants

6. Proper labeling, storage and use of toxic compounds

7. Control of employee health conditions that could result in the microbiological contamination of food, food packaging materials and food contact surfaces

8. Exclusion of pests from the food plant

Each processor shall maintain sanitation control records at a minimum, document showing that the monitoring and corrections carried out. Sanitation controls may be included in the HACCP plan. The seafood HACCP Regulation recommends ("should") that each processor have and implement written Sanitation Standard Operating Procedures (SSOP), and required ("shall") that they monitor the sanitary conditions and practices during processing. FDA included sanitation control procedures as an integral part of the seafood HACCP regulations to encourage processors to pay more attention to routine sanitary practices. FDA felt the additional controls were necessary because:

1. Sanitation practices directly affect the microbiological safety of seafood products that are not further cooked by the consumer, such as ready – to – eat products, smoked products, raw molluscan shellfish, and other fish that are consumed raw

2. Sanitation practices are relevant to the microbiological safety of seafood products even where these products are to be cooked by the consumer; sanitation practices directly affect the chemical and physical safety of seafood products

3. Nearly half of the consumer complaints that FDA receives relating to seafood are related to the processing plant of food hygiene

4. Inspections conducted by FDA and the National Marine Fisheries Service (NMFS) demonstrate that a significant portion of seafood processors operate under poor sanitation conditions

The new mandated sanitation control procedures focus on specific parts of the GMP. They introduce new requirements for monitoring, corrections, and records keeping that are not specified in the GMP. The HACCP-like features for monitoring and record keeping were considered necessary "to develop a culture throughout the seafood industry in which processors assume an operative role in controlling sanitation in their plants". The sanitation control procedures and GMP together form the foundation for a complete seafood safety programme topped with a HACCP programme that is product and process specific. HACCP cannot succeed in a plant that does not have adequate GMP.

HACCP vs. SCPs

A complete food safety programme includes both a HACCP plan and the accompanying sanitation control procedures based on GMP. Both components require monitoring, corrections and record keeping; yet there are a few regulatory subtleties that should be distinguished for each component. Distinction between HACCP and SCP is not always clear.

Difference between HACCP and SCPs

Hazard	Control	Type of Control	Control Programme
Histamine	Time & temperature of scombroid fish	Product specific	CCP
Pathogen Survival	Time & temperature for smoking fish	Processing step	CCP
Contamination with pathogens	Wash hands before touching product	Personnel	Sanitation
Contamination with pathogens	Limit employee movement between raw and cooked areas	Personnel	Sanitation
Contamination with pathogens	Clean and sanitize food contact surfaces	Plant environment	Sanitation
Chemical contamination	Use only food-grade grease	Plant environment	Sanitation

Recommended level of available chlorine in water used for seafood processing

Stage of processing	Recommended level of available chlorine
For water used in the processing plant	5 - 10 ppm
For ice manufacture	5 - 10 ppm
For washing contaminated ice before using with the fish	5 - 10 ppm
For disinfecting the fish processing factories and primary processing centres after applying a suitable detergent (contact time - 15 min.)	100 ppm
For disinfecting floor surface, gutters, etc.	500 - 800 ppm
For final washing	5 - 10 ppm
For disinfecting boat decks, fish holds, wooden boxes, etc. (contact time -5 min.)	1000 ppm
For spraying fish containers, fish carrier vans and refrigerated wagons in order to remove fish smell	100 ppm
For disinfecting washed utensils coming in contact with seafood (immersion time -5 min.)	100 ppm
For disinfecting workers' washed hands	200 ppm
For cooling cooked frozen prawns	20 ppm
For dipping the material before packing (immersion time - 10 min.)	20 ppm
For glazing cooked frozen prawns	10 ppm
For re-glazing cooked frozen prawns	20 ppm
For cooling processed cans	3 - 5 ppm

7.3. GOOD HYGIENIC PRACTICES (GHP)

The cultural attitudes of humans in food industries should not be allowed to compromise public health safety. Food handlers can cause contamination at every stage of the food chain from harvest to final consumption and cause consumers to become ill. Food-borne diseases are largely preventable, if good personal hygiene is observed. All personnel working with food should be trained in basic food hygiene. In many instances, the food handlers must provide a medical certificate. In all cases, cuts, wounds or abrasions must be covered with a waterproof dressing, and workers suffering from any communicable diseases (including hepatitis, diarrhoea, vomiting, ear, nose and throat illnesses, and boils) should not be allowed in the food processing establishments. Clean protective clothing should be provided and worn only in the processing facility. This includes overalls or coats (with no external pockets), footwear and head covering. Where RTE products are processed, face masks may be required. Hair should be kept covered and hair fasteners should not be used. Jewellery should not be worn. Smoking, spitting or eating in processing areas is not permitted. Work place should be cleaned after use.

Good hygienic practices include the following:

- Provision of safe water and ice
- Cleanliness of food contact surfaces
- Prevention of cross contamination from insanitary objects to food
- Maintenance of facilities for personal hygiene
- Protection of food from adulterants
- Safe storage and use of toxic compounds
- Provision of pest control
- Proper waste management
- Clean transportation

The operational hygiene requirement for safe seafood are given below:

Water

- Adequate supply of potable water with sufficient pressure and volume
- Automatic treatment system adapted and operational
- Clear distinctions between potable and non-potable water pipes
- Water quality monitored regularly
- Monitoring of residual chlorine content, if added
- Surveillance of contamination indicators in place.

Sampling plan is adequate and systematically followed.

Ice

- Ice is produced from potable water / clean water
- Ice stored in clean and well-maintained containers designated for this purpose
- Safety of ice is monitored

Steam

- Steam in contact with fish and shellfish made from potable water
- Steam available at sufficient pressure

Staff facilities

- Adequate changing facilities with separate changing rooms for men and women in different processing areas
- Sufficient flush toilets that are connected to an effective drainage system
- Lavatories located away from production, packing and storage areas
- Adequate number of washbasins for cleaning hands provided with running water and non-hand–operated water taps, and materials for cleaning hands and for hygienic drying
- Staff facilities area properly maintained and kept clean.

Hygiene control programme

- Appropriate cleaning and disinfection plan implemented by trained workers
- Persons who use physical, chemical and biological means for cleaning and disinfection properly trained

Waste management

- Offal and other waste regularly removed from production areas, so that no accumulation occurs.
- Sufficient closable containers for offal and other waste, clearly identified and made of easy-to-clean and impervious materials with suitable structure

- Adequate provision is made for the storage and disposal of food waste and waste materials
- Waste stores is designed and managed in such a way as to enable easy cleaning, and to prevent ingress of animals and other pests
- Drainage channels are designed to ensure that waste does not flow from a contaminated area towards or into a clean area
- All waste is eliminated in a hygienic and environmentally friendly way, and is not constituting a direct or indirect source of contamination

Pest control systems

- Good hygienic practices employed throughout to avoid pest infestation
- Pest control programme is available that prevents access, eliminates harbourage and infestation, and establishes monitoring, detection and eradication system
- Physical, chemical and biological agents for pest control procedures are properly applied by qualified personnel
- Rodenticides, insecticides, disinfectants and pest control are properly applied by qualified personnel
- Rodenticides, insecticides, disinfectants and any other toxic substance stored in premises or cupboards that can be locked
- Toxic products cannot contaminate the fish products

Personal hygiene and health

- Persons working in a fish handling area maintain a high degree of personal hygiene
- All people entering the area where fish is handled are provided with suitable, clean and protective clothing (uniform, aprons, rubber boots, gloves, hairnet)
- Protective clothing is cleaned by the company
- Medical examination is periodically carried out on workers handling fish
- Workers who could contaminate the products are excluded from handling fish and fishery products

- Workers handling fish wash and disinfect their hands each time before resuming work
- Workers keep their fingernails short, clean and unvarnished
- Any wounds is covered with waterproof bandages
- Smoking, spitting and eating are prohibited in production, packaging and storage areas
- Workers are trained in and follow the hygiene instructions
- First-aid assistance or first- aid cabinet is available
- Medical personnel are available when factory is working

7.4 Good Manufacturing Practices (GMP)

The Good Manufacturing Practices (GMPs) are required for HACCP compliance of aquatic food and mainly focus on eight control steps pertaining to sanitation. GMP is the term recognized worldwide for the control and management of production and quality control of foods. As per the WHO, the GMP is that part of the quality assurance, which ensures that the products are consistently produced and controlled to the quality standards appropriate to their indented use and is required by the marketing authorization. The GMPs are necessary for the production of quality and safe aquatic food. It ensures that quality and safety are maintained throughout a process and thus prevents product rejection and financial losses. The GMPs deal with contamination by people, food materials, packaging materials, hazardous materials, and miscellaneous materials. The GMP covers raw materials, purchasing and control, process control, premises, personnel, final product, and distribution. The GMPs are the provision of safe water for processing of food, usage of clean food contact surfaces, prevention of cross-contamination, maintenance of employee health, enforcement of employee sanitation, protection against adulterant, usage of correct pest control practices, and proper labelling, storage and usage of toxic compounds meant for cleaning and control of pests in the food industry. The principles of GMP are:

1. Design and proper construction of facilities and equipments
2. Follow written procedures and instructions
3. Document work
4. Validate work

5. Monitor facilities and equipments
6. Write step by step operating procedures & work-on instructions
7. Design, develop and demonstrate job competence
8. Protect against contamination
9. Control components and product related process
10. Conduct plan and periodic audits

The Current Good Manufacturing Practices (cGMPs) are the guidelines that ensure the right designs, monitoring procedures and the required control of the manufacturing processes and the facilities used. Compliance to the cGMP assures acquisition of identity, purity and high quality standards of the food commodities by forcing the manufacturers to apply the right control measures.

The GMP can be considered as a traditional method of quality assurance. The cGMP is based upon the same principles but it adheres to the standards along with latest available technology. Hence, it can be considered as a more effective and efficient method of ensuring quality assurance. Due to this reason, manufacturers should think about following cGMP instead of GMP.

CHAPTER 8

——————•——————

STANDARDS FOR AQUATIC FOODS

Standard refers to reasonably complete and widely applied food product specification that has been agreed nationally or internationally. It is also a technical specification or other document available to the public, drawn up with the cooperation and consensus or general approval of all interests affected by it, based on the consolidated results of science, technology, experience, aimed at the promotion of optimum community benefits and approved by a body recognized on the national or international level. Regulation is a binding document, which contains legislative, regulatory or administrative rules that is adopted and published by an authority legally vested with the necessary powers. Technical specification is a document, which lays down the characteristics of a product or a service such as levels of quality, performance, safety and dimension. It may include terminology, symbols, testing and test methods, packaging and labelling requirements. Code of practice is a processing specification that applies generally to processes of a given type. Codes are guidelines normally for a whole industry drawn up by regional, national or international agencies. They are almost invariably voluntary or advisory in nature. Some of their features relating to health protection have the force of law, as in the USFDA's Good Manufacturing Practices. Processing details are included in some mandatory standards. Industrial processing specification deals with the particular product or equipment. Standards have been introduced to protect the health of the consumers and to ensure fair practices in seafood trade. The formulation of standards for fish and fishery products are necessary to attain a minimum standard of cleanliness and hygiene in fish handling, processing and marketing. Standards are intended to guide and promote export or import of fishery products between countries. These standards fall into safety standards and composition standards. Safety standards are formulated to protect the consumer against food that are damaging the health. This ensures that reasonable standards of hygiene are practiced so that seafood are free of pathogens and that use of food additives are controlled and contaminants are prevented. Composition standards protect the consumer against fraud by ensuring that seafood is unadulterated, pure and of good quality. Packages should contain the correct description, labeling, weights, etc. Examples are fish pasta and fish pastes and fish fingers where the composition is printed on the label. In addition to these seafood laws, various national and international standards and codes of practices exist in order to place good quality and safe products in the market.

8.1. NATIONAL STANDARDS

Many fish producing countries have their own standards and codes of practice for fishery products. The Food Safety and Standards Authority of India (KSSAI) Indian Standards Institution ISI (renamed Bureau of Indian Standards, BIS), British Standards, BS, United States Food and Drug Administration, etc. have brought out standards for their fish and fishery products. They govern the quality and standards of products (fish / fishery) for local consumption as well as those for export and import.

8.1.1. Food Safety and Standards Authority of India (FSSAI)

The Food Safety and Standards Authority of India (FSSAI) was established under Food Safety and Standards, 2006 which consolidates various Acts & Orders that have hitherto handled food related issues in various Ministries and Departments. FSSAI has been created for laying down science based standards for articles of food and to regulate their manufacture, storage, distribution, sale and import to ensure availability of safe and wholesome food for human consumption.

Highlights of the Food Safety and Standards Act, 2006

- Various Central Acts like Prevention of Food Adulteration Act,1954, Fruit Products Order, 1955, Meat Food Products Order, 1973
- Vegetable Oil Products (Control) Order, 1947, Edible Oils Packaging (Regulation) Order, 1988, Solvent Extracted Oil, De-Oiled Meal and Edible Flour (Control) Order, 1967, Milk and Milk Products Order, 1992, etc. will be repealed after commencement of FSS Act, 2006.

The Act also aims to establish a single reference point for all matters relating to food safety and standards, by moving from multi-level, multi- departmental control to a single line of command. To this effect, the Act establishes an independent statutory Authority – the Food Safety and Standards Authority of India with head office at New Delhi. Food Safety and Standards Authority of India (FSSAI) and the State Food Safety Authorities shall enforce various provisions of the Act.

Establishment of the Authority

Ministry of Health & Family Welfare, Government of India is the Administrative Ministry for the implementation of FSSAI. The Chairperson and Chief Executive Officer of Food Safety and Standards Authority of India (FSSAI) have already been appointed by Government of India. The Chairperson is in the rank of Secretary to Government of India.

Functions of FSSAI

- Framing of Regulations to lay down the standards and guidelines in relation to articles of food and specifying appropriate system for enforcing various standards thus notified
- Laying down mechanisms and guidelines for accreditation of certification bodies engaged in certification of food safety management system for food businesses
- Laying down procedure and guidelines for accreditation of laboratories and notification of the accredited laboratories

- Providing scientific advice and technical support to Central Government and State Governments in the matters of framing the policy and rules in areas, which have a direct or indirect bearing on food safety and nutrition

- Collecting and collating data regarding food consumption, incidence and prevalence of biological risk, contaminants in food, residues of various contaminants in foods products, identification of emerging risks and introduction of rapid alert system

- Creating an information network across the country so that the public, consumers, panchayats. etc receive rapid, reliable and objective information about food safety and issues of concern.

- Providing training programmes for persons who are involved or intended to get involved in food businesses

- Contributing to the development of international technical standards for food, sanitary and phytosanitary

- Promoting general awareness about food safety and food standards

Food Authority

As per Section 5 of the FSS Act, the Food Authority shall consist of a Chairperson and the following twenty-two members out of which one-third shall be women, namely:-

(a) seven Members, not below the rank of a Joint Secretary to the Government of India, to be appointed by the Central Government, to respectively represent the Ministries or Departments of the Central Government dealing with Agriculture, Commerce, Consumer Affairs, Food Processing, Health, Legislative Affairs, and Small Scale Industries, who shall be Members ex officio

(b) two representatives from food industry of which one shall be from small scale industries

(c) two representatives from consumer organisations

(d) three eminent food technologists or scientists

(e) five members to be appointed by rotation every three years, one each in seriatim from the Zones as specified in the First Schedule to represent the States and the Union territories

(f) two persons to represent farmers' organisations

(g) one person to represent retailers' organisations

Scientific Committee

The Chairpersons of all Scientific Panels, along with six experts nominated by FSSAI shall form the Scientific Committee.

Scientific Panels

There are 17 Scientific Panels formed by FSSAI, which are given below:

- Panel on Food additives, Flavourings, Processing aids and Materials in contact with food
- Panel on Methods of Sampling and Analysis
- Panel on Biological Hazards
- Panel for Contaminants in the Food Chain
- Panel for Pesticides and Antibiotic Residues
- Panel on Labelling and Claims/Advertisements
- Panel on Functional Foods, Nutraceuticals, Dietetic Products and Other similar products
- Panel on Fish and Fisheries Products
- Panel on Genetically Modified Organisms and Foods
- Panel on Nutrition and Fortification
- Panel on Milk & Milk Products
- Panel on Meat & Meat Products including poultry
- Panel on Cereals, Pulses & Legume and their products (including Bakery)
- Panel on Fruits & Vegetables and their products (including dried fruits and nuts salt, spices and condiments)
- Panel on Oils & Fats
- Panel on Sweets, Confectionery, Sweeteners, Sugar & Honey
- Panel on Water (including flavoured water) & Beverages (alcoholic and non-alcoholic)

The Panel on Fish and Fisheries Products has 11 expert members from Central/ State Research Organizations/ Institutes and Universities.

Risk Assessment

The FSSAI under Section10, 16 (1)(i) (c) and 18(1) (2) (b)(c), has established a Risk Assessment Cell (RAC) to improve Food Safety Frameworks. The RAC will carry out functions of risk assessment to support risk management and risk communication. Broadly, a risk assessment will be done for products, processes and activities that could result in an increase in a health risk and for anything that might have a direct effect on food safety. The RAC will provide an effective framework for determining the relative urgency of problems related to food safety and the remedial measures to reduce or control these risks. It will be working on identification of the hazards that cause food borne health problems, in a variety of ways including, the data collection on contaminants and adulterants in food categories through laboratories, research & development projects (Pull and Push type), lack of compliance data with regards to standards, labelling and other issues. Once the areas of concern are identified, it justifies the need for risk analysis. This will later help in developing and improving the present food safety system including the regulations, developing science based standards, food safety policies, laboratory analysis, and epidemiological surveillance. The food hazards identified can be numerous and all cannot be addressed at the same time by the Authority, RAC will be working on ranking the hazards, to determine which area of concern needs to be addressed first and establishing the priorities, based on public health as well as economic and social aspect of the risk, for management. The aim should be how to reduce the risk, protect the trades associated with the risk and protect international trade practices.

Objectives of Risk Assessment Cell (RAC)

1. To carry out the basic work of risk assessment through collection, processing and analysis of surveillance data and prepare an analysis report.
2. Provide inputs to the Scientific Panels and Scientific Committee for review and risk assessment outputs and scientific opinion to assist in management of food safety hazards in the areas of concern.

3. Provide and utilize a pool of scientific expertise across the country through Food Safety Knowledge Assimilation Network (FSKAN) for scientific and technical opinions

4. Communicate the risk assessment findings conducted by Scientific Panels and Scientific Committee to the risk managers and other stakeholders

5. Coordinate in setting vertical and horizontal standards for food products

Methodology (Risk Assessment process)

Risk assessments are resource-intensive, data-driven activities designed to provide risk managers with the ability to consider a range of mitigations that are intended to improve public health. There is no one way to perform the food safety risk assessment and it varies according to the nature of the risk, the availability of data and other scientific evidences to answer specific risk management queries. Currently, efforts are being made for pooling/ sourcing of scientific data from research institutions, export agencies, and other government organizations at national level. The data collected would be helpful in carrying out risk assessment and surveillance plans.

Work flow

1. After identifying and expressing the nature and characteristics of the food safety issues, Risk profile is developed by the RAC which will be describing the situation, product or commodity involved and the information on the pathways by which the consumers are exposed to the hazard.

2. The decision weather a risk assessment is necessary will be taken jointly by the Committee comprising of experts and the FSSAI officials

3. A request proposal for commissioning of risk assessment will be issued by the RAC for the relevant Scientific Panel

4. The data from the risk assessment studies will be used as an inputs to be provided to the relevant scientific panel for scientific opinions, scientific opinion provided by the Panels will be either vetted by the scientific committee for further decision. If approved, it will be forwarded to Food Regulators and Food Authority for policy decisions.

5. FSSAI and Food Authority will provide decision /options to mitigate or control the risk.

6. In case of any further clarifications on the risk assessment output/ scientific opinion from Scientific committee or Food Authority the RAC will coordinate and provide the same for consideration in subsequent meetings.

7. FSSAI, as risk Managers, will be responsible for considering the scientific opinion for decision making process and in developing a risk management policy.

8. FSSAI will also be responsible for risk communication portion, this being a two-way process and involves sharing the information internally between the risk analysis team and with external stakeholders including general public.

9. Essential database required to conduct risk assessment will be developed and maintained by RAC.

Food Safety and Standards Regulations 2011

Section 92 (1) of Food Safety and Standards Act, 2006 empowers the Food Authority to make regulations/standards in consistent with this Act and Rules made thereunder. After the enactment of the FSS Act 2006, the FSSAI has drafted six principal regulations through extensive consultation and deliberations/ meetings with various stakeholders. These regulations have been notified in the Gazette of India on 1st August, 2011 and came into force on 5th August, 2011. These regulations/standards are further reviewed and several new regulations are notified taking into account the latest developments in food science, food consumption pattern, new products and additives, advancement in the processing technology and food analytical methods and also with a view to bridge the gap between national and international standards for food products. The formulation and revision of regulations/ standards involves several stages. After recommendation by the Scientific Panel, following the due process laid down including validation by the Scientific Committee, a regulation/ standard is approved by the Food Authority. After soliciting comments of stakeholders and their considerations, final notification is issued for implementation.

In addition to the six principal regulations notified on 1st August 2011, the following new regulations were notified. These include:

1. Food Safety and Standards (Food or Health Supplements, Nutraceuticals, Foods for Special Dietary Uses, Foods for Special Medical Purpose, Functional Foods and Novel Food) Regulation, 2016
2. Food Safety and Standards (Food Recall Procedure) Regulation, 2017
3. Food Safety and Standards (Import) Regulation, 2017
4. Food Safety and Standards (Approval for Non-Specified Food and Food Ingredients) Regulations, 2017
5. Food Safety and Standards (Organic Food) Regulation, 2017

FSSAI Gazette Notification on Microbiological Standards for Fish and Fisheries Products

On 13 February, 2017, the FSSAI issued a final Gazette Notification on Food Safety and Standards (Food Product Standards and Food Additives) Third Amendment Regulation, 2017 related to microbiological standards for fish and fisheries products. These regulations came into force on 1st January 2018.

In the microbiological requirements for fish and fishery products under hygiene indicator organisms, the sampling plan and limits are given for Aerobic Plate Count, Coagulase-positive Staphylococci, and Yeast and mould count. In the microbiological requirements for fish and fishery products under safety indicator organisms, the sampling plan and limits for the fish products have been given for microorganisms like *Escherichia coli, Salmonella, Vibrio cholerae* (O1 and O139), *Listeria monocytogenes* and *Clostridium botulinum.*

The microbiological standards are applicable to fish and fishery products such as Chilled/Frozen Finfish, Chilled/Frozen Crustaceans, Chilled/Frozen Cephalopods, Live Bivalve Molluscs, Chilled/Frozen Bivalves, Frozen cooked Crustaceans or Frozen heat shucked Mollusca, Dried or Salted and Dried fishery Products, Thermally Processed Fishery Products, Fermented Fishery Products, Smoked Fishery Products, Accelerated Freeze-dried Fishery Products, Fish Mince/ Surimi and Surimi Analogues, Fish Pickle, Battered and Breaded Fishery Products, Convenience Fishery Products, and Powdered Fish based Products.

Chilled/Frozen Finfish include clean and wholesome finfish which can be raw, chilled or frozen and handled in accordance with good manufacturing practices. Chilling can be achieved either by using ice, chilled water, ice slurries of both seawater and freshwater or refrigerated seawater. Freezing preserves the inherent quality of the fish during transportation, storage and distribution up to and including the time of final sale by maintaining low temperature in appropriate equipment in such a way that the range of temperature of the product must reach −18°C or lower at the thermal centre after thermal stabilization.

Chilled/Frozen Crustaceans include clean, whole or peeled crustaceans like shrimp, prawn, crabs and lobster which can be raw, chilled or frozen and handled in accordance with good manufacturing practices.

Chilled/Frozen Cephalopods include cleaned, whole or de-skinned cephalopods like squid, cuttlefish and octopus and can be raw, chilled or frozen and handled in accordance with good manufacturing practices.

Live Bivalve Molluscs include oyster, clam, mussel, scallop and abalone which are alive immediately prior to consumption and are presented in the shell. Live bivalve molluscs are harvested alive from a harvesting area either approved for direct human consumption or classified to permit harvesting for an approved method of purification, like depuration, prior to human consumption and both types are subject to appropriate controls implemented by the official agency having jurisdiction.

Chilled/Frozen Bivalves includes clean, whole or shucked bivalves, which are live either in chilled or frozen condition and handled in accordance with good manufacturing practices. This product category includes filter feeding aquatic animals such as oysters, mussels, clams, cockles and scallops.

Frozen cooked Crustaceans or Frozen heat shucked Mollusca means clean, whole or peeled crustaceans, which are cooked at a defined temperature and time and subsequently frozen. Cooking of crustaceans must be designed to eliminate most heat resistant vegetative bacteria i.e. *Listeria monocytogenes*. Frozen heat shucked mollusca includes bivalves where meat is removed from the shell by subjecting the animals to mild heat before shucking to relax the adductor muscle and subsequently frozen.

Dried or Salted and Dried fishery Products mean the product prepared from fresh or wholesome finfish or shellfish after drying with or without addition of salt. The fish shall be bled, gutted, beheaded, split or filleted and washed prior to salting and drying.

Thermally Processed Fishery Products mean the product obtained by application of heat or temperature for sufficient time to achieve commercial sterility in hermetically sealed containers.

Fermented Fishery Products include any fish product that has undergone degradative changes through enzymatic or microbiological activity either in presence or absence of salt. Products manufactured by accelerated fermentation, acid ensilage and chemical hydrolysis also belong to this category.

Smoked Fishery Products mean fish or fishery products subjected to a process of treatment with smoke generated from smouldering wood or plant materials. Here, the product category refers to hot smoked fish where fish is smoked at an appropriate combination of temperature and time sufficient to cause the complete coagulation of the proteins in the fish flesh.

Accelerated Freeze-dried Fishery Products mean fish, shrimp or any fishery product subjected to rapid freezing, followed by drying under high vacuum so as to remove the water by sublimation to final moisture content of less than 2%.

Fish Mince/Surimi and Surimi Analogues mean comminute, mechanically removed meat which has been separated from and are essentially free from bones, viscera and skin. Surimi is the stabilized myofibrillar proteins obtained from mechanically deboned fish flesh that is washed with water and blended with cryoprotectants. Surimi analogues are variety of imitation products produced from surimi with addition of ingredients and flavour.

Fish Pickle means an oily, semi-solid product with spices and acidic taste obtained from maturation of partially fried fish with vinegar. It is produced by frying edible portions of fish, shrimp or mollusc, followed by partial cooking with spices, salt and oil and maturing for 1-3 days with added organic acids. The product is intended for direct human consumption as a seasoning, or condiment for food.

Battered and Breaded Fishery Products include fish portions, fillets or mince coated with batter or breading or both. Batter means liquid preparation from ground cereals, spices, salt, sugar and other ingredients and/or additives for coating. Typical batter types are non-leavened batter and leavened batter. Breading means dry bread crumbs or other dry preparations mainly from cereals with colourants and other ingredients used for the final coating of fishery products.

Convenience Fishery Products are tertiary food products made of fish, which are in ready to eat form and also includes snack based items prepared from fish and fishery products meant for direct human consumption such as extruded fishery products, fried items namely fish wafers, crackers, fish cutlets, fish burgers and other such products. These products can be consumed directly after minimal handling and processing. This category includes Sous vide cooked products, surimi based products cooked (in pack), pasteurized crab meat, pasteurized molluscs which are distributed as refrigerated, but meant for direct human consumption with minimal or no cooking.

Powdered Fish based Products include the products which are prepared from finfish/shellfish or parts thereof, with or without other edible ingredients in powdered form, suitable for human consumption. These may be consumed directly or as supplements and also after hydration and this category includes powdered and dried fish products generally used as ingredients in food preparations such as fish soup powder, fish chutney powder, ready to use fish-mix, and such other food.

FSSAI Gazette Notification on Vertical Standards for Fish and Fisheries Products

The FSSAI has notified the final Food Safety and Standards (Food Products Standards and Food Additives) Eleventh Amendment Regulations, 2017 on 13.09.2017. Through this notification, the standards for frozen shrimps and frozen finfish and frozen fish fillets have been revised and comprehensive standard for canned fishery products have also been prescribed. This notification also prescribes the standards for frozen cephalopods, smoked fishery products, ready-to-eat finfish/ shellfish curry in retortable pouches, sardine oil, edible fish powder, fish pickle, frozen minced fish meat, freeze dried prawns, and frozen clam meat.

8.1.2. Bureau of Indian Standards (BIS)

The Bureau of Indian Standards (BIS), the erstwhile Indian Standards Institution, which started functioning in 1947, is the National Standards Organization in India. Its principal objective is to prepare standards on national basis and promote their adoption. The Bureau has brought out over fifty standards for various fish and fishery products. These standards prescribe detailed requirements of processing, packaging and methods of analysis for evaluation of quality of the fishery product. Salient features include scope of standards, terminology, grades, preparation of material, requirements, packaging and marketing, sampling and tests. Requirements of products are:

1. Physical aspects   -   Weight, size, etc.
2. Sensory aspects   -   Appearance, colour, texture, odour and flavour
3. Microbiological requirements   -   Total plate count, *E. coli*, Staphylococci, Streptococci and *Salmonella*
4. Chemical aspects   -   Moisture content, ash, sodium chloride, heavy metals

A. Fresh Fish

Under fresh fish, standards are available for pomfrets, mackerel, threadfin, seerfish, etc. The formulation of standards has been felt necessary with a view to make available fresh fish of desired quality and selecting raw material for freezing and canning purposes. It should be handled and transported under sanitary conditions, washed in water containing 5-10 ppm chlorine, precooled and iced. The fresh fish shall be clean, wholesome and fresh. It should have characteristics colour, odour, bright eyes, bright red gill, firm flesh, etc. The material may also satisfy the microbiological limits.

B. Frozen fish

Indian standards are available for frozen products like shrimp, lobster tails, crabmeat, cuttlefish, squid, pomfret, threadfin, mackerel, seerfish, etc. The main quality requirements of the frozen products are dehydration, drained weight, size grade, discoloration, decomposition and microbiological requirements. On thawing, the product should be clean, intact, undamaged condition and free from defects. Deteriorations such as dehydration, oxidative rancidity and adverse changes in the texture shall not be present. The products shall be free from foreign matter.

C. Dried and cured fish products

Under this, standards are available for dried prawns, dried white baits, dried and laminated Bombay duck, dry salted products like mackerel, seerfish, shark, tuna, threadfin, jewfish, catfish, horse mackerel, etc. The main quality requirements include material shall have characteristics dry-salted fish odour and shall not show red or pink discoloration, shall be free from off-odour indicative of spoilage, free from foreign matter, free from excessive sand and salt and free from insects and mite infestation and from visible fungal growth. The moisture, salt and ash contents are also prescribed. The curing period is also specified. The fish while drying shall be protected against contamination from dirt, sand, flies and insects.

D. Others

Standards are available for a variety of canned products, fish meal, shark liver oil and sardine oil. In addition to these, standards are also avialable on code of practices such as Code for hygienic conditions for fish industry (I – Preprocessing stage and II – Processing stage), Recommendations for maintenance of cleanliness in fish industry, Basic requirements for fresh fish stalls, Requirements for fish market, Procedure for checking temperature of quick frozen foods, Specification for master cartons for export of fishery products, and Methods of tests to achieve the various quality standards.

Indian Standards on Fish and Fishery Products

Sl. No.	Name of specification	Specification Number and Year of publication
	Fresh Fish	
1.	Fresh silver pomfret and brown pomfret	IS 4780 - 1968
2.	Fresh threadfin	IS 4781 - 1968
3.	Mackerel, fresh	IS 6032 - 1971
4.	Seerfish (*Scomberomorus* spp.), fresh	IS 6123 - 1971
	Frozen fish and shellfish	
5.	Frozen prawns (shrimp) - second revision	IS 2237 - 1985
6.	Frozen froglegs	IS 2885 - 1964
7.	Frozen lobster tails	IS 3892 - 1966
8.	Frozen threadfin	IS 4796 - 1968
9.	Frozen silver pomfrets and brown pomfrets	IS 4793 - 1968
10.	Mackerel, frozen	IS 6033 - 1971
11.	Seerfish (*Scomberomorus* spp.), frozen	IS 6122 - 1971
12.	Frozen cuttlefish and squid	IS 8076 - 1976
13.	Frozen minced fish meat	IS 10763 - 1983
	Canned fish and shellfish	
14.	Pomfret canned in oil (first revision)	IS 2168 - 1971
15.	Prawns (shrimp) canned in brine (first revision)	IS 2236 - 1968
16.	Mackerel (*Rastrelliger* spp.) canned in oil (second revision)	IS 2420 - 1985
17.	Mackerel (*Rastrelliger* spp.) canned in brine	IS 3849 - 1966
18.	Sardines (*Sardinella* spp.) canned in oil (first revision)	IS 2421 - 1971
19.	Sardines (*Sardinella* spp.) canned in brine and in their juice	IS 6677 - 1972
20.	*Lactarius* spp. canned in oil (first revision)	IS 6121 - 1985
21.	Tuna canned in oil	IS 4304 - 1967
22.	Crabmeat, canned in brine	IS 7143 - 1973
23.	Crabmeat solid packed	IS 7582 - 1975
24.	Mackerel canned in curry	IS 9312 - 1979
25.	Specification for mussel canned in oil	IS 10760 - 1983

[Table Contd.

Contd. Table]

Sl. No.	Name of specification	Specification Number and Year of publication
	Dried fish and shellfish	
26.	Dried prawns (second revision)	IS 2345 - 1985
27.	Dried and laminated Bombay duck	IS 2884 - 1964
28.	Dried white baits (Anchoviella spp.)	IS 2883 - 1985
29.	Dry salted mackerel	IS 4302 - 1967
30.	Dry salted seerfish (first revision)	IS 5198 - 1985
31.	Dry salted shark (first revision)	IS 5199 - 1985
32.	Dry salted surai (Tuna) (first revision)	IS 5736 - 1985
33.	Dry salted threadfin (Dara) and dry salted Jew fish (Ghol) first revision)	IS 3850 - 1973
34.	Dry salted cat fish	IS 3851 - 1966
35.	Dry salted leather jacket (*Chorinemus* spp.)	IS 3852 - 1966
36.	Dry salted horse mackerel (*Caranx* spp.)	IS 3852 - 1985
37.	Dry shark fin	IS 5471 - 1969
38.	Fish maws (first revision)	IS 5472 - 1985
39.	Dry salted dhoma (first revision)	IS 8836 - 1985
	Miscellaneous	
40.	Code for hygienic conditions for fish industry Part I Pre-processing stage (first revision)	IS 4303 - 1975
41.	Code of hygienic conditions for fish industry Part II Canning stage (first revision)	IS 4303 - 1975
42.	Recommendation for maintenance of cleanliness in fish industry	IS 5375 - 1970
43.	Fish meal as livestock feed (first revision)	IS 4307 - 1973
44.	Shark liver oil for veterinary use	IS 3336 - 1965
45.	Sardine oil	IS 5734 - 1970
46.	Glossary of important fish species of India	IS 7313 - 1974
47.	Basic requirements for fresh fish stalls	IS 7581 - 1975
48.	Basic requirements for fish market	IS 8082 - 1976
49.	Procedure for checking temperature of quick frozen foods	IS 8077 - 1976
50.	Methods for sampling fish and fishery products	IS 11427 - 1985

Microbiological requirements for fresh and frozen fish and shellfish (bacterial count maximum / g)

Sl.No.	Name of fish/shellfish	Fresh/Frozen	SPC	E. coli	Coagulase positive Staphylococci	Faecal Streptococci	Salmonella
1.	Mackerel	Fresh	1,00,000	20	–	–	Nil
2.	Threadfin	Fresh	1,00,000	20	–	–	"
3.	Pomfrets	Fresh	1,00,000	20	–	–	"
4.	Mackerel	Frozen	1,00,000	10	–	–	"
5.	Threadfin	Frozen	1,00,000	10	–	–	"
6.	Pomfrets	Frozen	1,00,000	10	–	–	"
7.	Seer fish	Frozen	1,00,000	10	–	–	"
8.	Lobster tails	Frozen	5,00,000	20	100	–	"
9.	Prawns (Whole & headless)	Frozen	5,00,000	20	100	100	"
10.	Prawns (Peeled & deveined)	Frozen	10,00,000	20	100	100	"
11.	Prawns (cooked)	Frozen	1,00,000	Nil	100	100	"
12.	Cuttlefish	Frozen	1,00,000	10	100	–	"

**Standards for water used in processed food industry (as per IS 4251)
and ice manufacture (as per IS 3957)**

Characteristics	Tolerance	
	Food Industry	Ice manufacture
1. Colour (Hazen units), Max	20	5
2. Turbidity (Units), Max	10	5
3. Odour	None	None
4. pH	6.5 to 9.2	6.5 to 9.2
5. Total dissolved solids, mg / L, Max.	1000	1000
6. Alkalinity (as $CaCO_3$), mg / L. Max.	–	100
7. Total hardness (as $CaCO_3$), mg / L. Max.	600	600
8. Sulphate (as SO_4), mg / L. Max.	200	200
9. Fluoride (as F), mg / L. Max.	1.5	1.5
10. Chloride (as Cl), mg / L. Max.	250	250
11. Cyanide (as CN), mg / L. Max.	0.01	0.01
12. Selenium (as Se), mg / L. Max.	0.05	0.05
13. Iron (as Fe), mg / L. Max.	0.3	0.3
14. Magnesium (as Mg), mg / L. Max.	75.0	125
15. Manganese (as Mn), mg / L. Max.	0.2	0.2
16. Copper (as Cu), mg / L. Max.	1.0	1.0
17. Lead (as Pb), mg / L. Max.	0.1	0.1
18. Chromium (as Cr+6), mg / L. Max.	0.05	0.05
19. Zinc (as Zn), mg / L. Max.	15.0	15.0
20. Arsenic (as As), mg / L. Max.	0.2	0.2
21. Nitrate (as N), mg / L. Max.	20	–
22. Phenolic substances (as C_6H_5OH), mg /L. Max.	0.001	0.001

8.1.3. Export Inspection Agency (EIA)

Govt. of India launched a comprehensive programme of quality control and pre-shipment inspection under the Export (Quality Control & Inspection) Act, 1963 in order to create an image of quality of Indian goods in the foreign markets and to promote exports on a long term and permanent basis. This Act, which came into force from 1st January 1964, empowers the Govt. of India to notify commodities, which shall be subject to compulsory quality control or inspection or both, before shipment. No consignment of the commodities so notified can be exported unless a certificate or the article accompanies it or package carries a recognized mark indicating that it conforms to the recognized specifications. The Act has provisions for the establishment of Export Inspection Council (EIC) to advise the Government in regard to the measures to be taken for quality control and pre-shipment of exportable commodities, such as standard specifications to be recognized, details of quality control and inspection methods and other related matters. EIC is the statutory body under the Ministry of Commerce of Govt. of India. The Council is also empowered to constitute specialist committees for conducting investigation on special problems connected with its functions.

The Govt. of India by notification in the official Gazette has established five Export Inspection Agencies (EIAs), one each at Bombay, Calcutta, Cochin, Delhi and Madras with effect from 1st February 1966, as the agencies for quality control and inspection of such commodities which are notified from time to time under the Act for the purpose of export by sea, land or air. All these five EIAs are under the administrative and technical control of EIC. Under the five EIAs, there are 66 sub-offices functioning in major ports and production centers of India for conducting inspection and testing. The Govt. of India has also recognized the following Institutions as Inspection Agencies under the Act. They are Agricultural Marketing Advisor to Govt. of India (AGMARK), Indian Standards Institution (ISI) and 51 Private Inspection Agencies. Initially about 40% of the commodities being exported from India are covered under the compulsory quality control and inspection schemes. EIC prescribes standards on organoleptical and microbiological quality characteristics of commodities. The main considerations for bringing a commodity under compulsory pre-shipment inspection are development of an export trade, volume and trend of export, extent of competition from other countries, and need for enforcing quality control or inspection or both for increasing the exports of the commodity

Compulsory quality control and pre-shipment inspection of fish and fishery products

The fish and fishery products (FFP), which were first brought under the compulsory quality control and pre-shipment inspection scheme are frozen shrimps and canned shrimps. At that time, the scheme was operated by the Director, Central Institute of Fisheries Technology, Cochin, who was recognized as the Inspection Agency under section 7 of the Act. The inspection scheme for FFP was taken over by EIA with effect from 1st May 1969. Gradually other items of FFP such as frozen frog legs, dried shark fins, fish maws, dried fish, dried prawns, frozen lobster tails, dried Bombay duck, canned crabmeat, frozen pomfrets, frozen cuttlefishes and squids, beche-de-mer and fish meal were also brought under the scheme. This scheme initially envisaged five types of inspection system to be followed in the export of fish and fishery products. They are consignment-wise (end product) inspection, in-process quality control (IPQC), modified in-process quality control (MIPQC), self-certification and voluntary inspection.

1. Consignment-wise inspection

In the beginning, the activities of the EIAs were mainly directed towards the inspection of the final products, which comprised drawal and inspection of samples on a random basis to ensure conformity of the product to the prescribed standard specifications. This inspection system can also be called as end product inspection. Dried fishery products are still inspected under this system.

2. In-process quality control (IPQC)

With a view to make the quality control & inspection (QC&I) scheme more meaningful and effective, the Govt. of India introduced the IPQC scheme for frozen shrimp, lobster, cuttlefish, squid and pomfret with effect from 31st January 1977. Canned fish and fishery products (prawns and crab meat) were brought under the IPQC scheme with effect from 12th February 1983. Under this scheme, processing of FFP for export is permitted only in the processing units, which are approved by EIA. To qualify for such approval, a production unit must have the minimum sanitary and hygienic facilities stipulated in the Export of Fish and Fishery Products (Quality Control & Inspection) Rules, 1977 or the Export of Canned Fishery Products (Quality Control & Inspection) Rules, 1983. These rules prescribe requirements in respect of the processing units, environmental, sanitary and hygienic aspects of the units, processing facilities available and health and personnel hygiene of the workers employed. As per the IPQC system, the entire processing operations are carried out under the supervision of a competent technical personal employed by the processing unit. Standards have also been laid down for the acceptance of raw material and only approved raw material is allowed for processing. At all the stages of processing, necessary inspection is conducted by the technologist of the unit. His/ her observations are regularly counterchecked by the Officer of EIA. The final product is further subjected to physical, organoleptical and bacteriological examinations and, if on such examinations, the product is found to be satisfactory, only then the certificate of export worthiness is issued by EIA for the relevant consignment. Subsequent to certification, EIA conducts further checks on the certified export consignments in the cold storages and at wharfs. Thus, IPQC scheme entails continuous surveillance and vigil by EIA Officers during the entire processing and packing operations starting with the receipt of the raw material till the final product is exported. This scheme was later renamed as Quality Control and Inspection in Approved units (QCIA) in 1987.

3. Modified in-process quality control (MIPQC)

The EIC felt that the system of IPQC might be modified in the case of those of the processing units, which have been accorded approval to process fishery products under the IPQC scheme and which have certain additional infrastructural facilities like their own testing laboratory to ensure production of bacteriologically and organoleptically sound products for export. Such units under the MIPQC system are permitted to process and pack fish and fishery products for export under their own supervision and control. The approved technologist of the processing unit conducts the in-process quality control drills, bacteriological examination of raw material and final product and organoleptical examination of the final product. The main duties of EIA Officers in respect of these units are to assist and guide the processors in their attempt to produce wholesome and quality products. EIA has provided detailed guidelines to the processors under MIPQC system on their duties and responsibilities under the scheme. Although the responsibility for ensuring the quality of the product and declaring the consignment exportworthy rests with the processors themselves, the inspection certificates are being issued by EIA for the purpose of export, with counter-checks wherever called for. This scheme was later referred to as In-process Quality Control (IPQC) in 1987.

4. Self-certification scheme (SCS)

The units under this scheme would be responsible to ensure the quality of their products as required by law and also to act as Inspection Agency to certify the cargo manufactured by them for export. The units that performed under IPQC for a stipulated period without any problems and rejections are eligible for self-certification. The system is extended to the processors having all infrastructural facilities, strict quality control system, regular quality auditing and established reputation. Exporters qualifying for SCS are listed in the official Gazette authorizing them to certify their own products.

5. Voluntary pre-shipment inspection

For the benefit of foreign buyers who would like to get the products inspected before shipment in respect of these commodities, which are not yet brought under the purview of compulsory pre-shipment inspection, arrangements are made for voluntary inspection.

Approval of Indian fish processing establishments by EU

The European Union (EU) in 1997 banned the import of Indian aquatic foods into EU countries due to the presence of pathogens in them. The Govt. of India took up the issue with the EU and after bilateral negotiations, a new system of approving the aquatic food processing plant was introduced. All processing factories exporting aquatic foods to EU countries are now expected to get approval as per EU norms 91/493/EEC. EIC continues to be the competent authority for quality assurance in this country as per Quality Control, Inspection and Monitoring Rules, 1995. According to this, the factories are first assessed by the Inter-Departmental Panel (IDP), a team of officials drawn from EIA, Central Institute of Fisheries Technology (CIFT) and Marine Products Export Development Authority (MPEDA). A team of Experts, called the Supervisory Audit Team (SAT) drawn from CIFT assesses the processing units cleared by IDP. Currently the EIA also utilizes the experts available in certain fisheries colleges across India for this purpose. Based on the final recommendation given by the SAT, the factory is approved for export to European Union. All the EU approved units are monitored by SAT once in 2 months. As per the system, the plant should have excellent hygienic pre-processing and processing facilities, ice production, cold storage, water purification system, etc. The movement of the product is unidirectional inside the plant. It should also have sufficient change rooms, toilet facilities and rest rooms, which depends on number of workers. The plant should implement the SSOP, GMP and HACCP with all relevant documents and monitoring procedures. The processing plants, which are not qualified for EU approval, are approved by EIA, based on the recommendation of panel of experts comprising one member each from EIA, MPEDA, CIFT and the industry, as nationally approved for exporting seafood to non-European countries.

8.2. INTERNATIONAL STANDARDS

8.2.1. ISO 9000

International Organization for Standards (ISO) was established in 1946 at Geneva, Switzerland with the aim of developing uniform international manufacturing trade and communication standards. The purpose of ISO is to promote the development of standardization and related activities in order to facilitate the international exchange of goods and services and to develop cooperation in intellectual, scientific, technological and economic activities. The ISO with 14 founding members from Europe, the US, and the British Commonwealth has grown to a worldwide federation with over 136 member countries, a decentralized pattern of work entrusted to over 180 active technical committees (TC) and over 620 active sub-committees, enlisting over 30,000 specialists engaged in developing international standards. ISO / TC – 176 was established in 1979 to focus on standards in quality management and quality assurance, which completed its task of development of the ISO core series of standards in 1987. Based on the good experience gained with the British Standards (BS) 5750 series published in 1979, they were adopted by ISO, and the ISO 9000 series were published in 1987 aiming at providing an international acknowledgement of quality efforts. More than 50 countries have now adopted the ISO 9000 series, which is equivalent to the BS 5750 standards. The European Community has adopted the ISO Standards as the European Norm (EN) 29000 series. In the case of United States, these standards are published by the American National Standards Institute (ANSI) and the American Society for Quality Control (ASQC) into the ANSI / ASQC Q 9000 series. Bureau of Indian Standards (BIS) has adopted the ISO 9000 series of standards as the Indian Standards (IS) 14000 series. The ISO 9000 standards are a series of five documents as shown below:

ISO Standards	Title
ISO 9000	Quality management and quality assurance standards –guidelines for selection and use
ISO 9000 – 1	Guidelines for selection and use
ISO 9000 – 2	Generic guidelines for the application of ISO 9001, ISO 9002 and ISO 9003
ISO 9000 – 3	Guidelines for the application of ISO 9001 to the development, supply and maintenance of software
ISO 9000 – 4	Guide to dependability programme management
ISO 9001	Quality systems – model for quality assurance design / development, production, installation and servicing
ISO 9002	Quality systems – model for quality assurance in production and installation
ISO 9003	Quality systems – model for quality assurance in final inspection and test
ISO 9004	Quality management and quality system elements – guidelines
ISO 9004 – 1	Guidelines
ISO 9004 – 2	Guidelines for services
ISO 9004 – 3	Guidelines for processed materials
ISO 9004 – 4	Guidelines for quality improvement
ISO 9004 – 5	Guidelines for quality plans
ISO 9004 – 6	Guidelines for project management
ISO 9004 – 7	Guidelines for configuration management

ISO standards are of two types in nature, which are conformance standards and guidance standards.

(a) Conformance standards

ISO 9001 -	Includes all elements in the promotion cycle from designing to servicing, and contains 20 requirements – design products
ISO 9002 -	Same as ISO 9001, except that there are no requirements for design control (manufacture only)
ISO 9003 -	For final inspection and test (neither design nor manufacture)

(b) Guidance standards

ISO 9000 -	It is guidance standard for selecting the proper conformance standard
ISO 9004 -	It gives details regarding quality management and quality system elements

The field of application of different ISO Standard are given below:

ISO Standard	Field of application
ISO 9000	Selection of the appropriate ISO 9000 standard
ISO 9001	Quality system requirements for product development, production, delivery and after sales functions
ISO 9002	Quality system requirements for production and delivery
ISO 9003	Quality system requirements for final inspection and testing
ISO 9004	Guidelines for ISO 9000, quality system elements

ISO 9001, 9002 and 9003 are three specific standards describing the elements and requirements of a quality system to be implemented in a company in connection with a contractual situation, i.e. the supplier – customer quality systems, and they form the background for obtaining a Quality System Certificate issued by an approved, independent organization or certifying body. ISO 9001 is the most comprehensive standard containing most of the elements described in the guidelines given in ISO 9004. Compared to ISO 9002, the most important difference is that it includes development of new products and processes. The ISO 9003 is used in situations where requirements to the producer comprising only final product inspection and testing and this standard only contains a minor part of the elements of ISO 9004. Most relevant standards for food processing companies are ISO 9001 and 9002. However, it shall be mentioned that the combined use of standards can be advantageous. For small enterprises like a fishing boat, one could use ISO 9003 and add the relevant elements of ISO 9002. This can lead to the most appropriate system manageable by such a small entity. In such a case, the official certification is according to the ISO 9003.

Elements of the quality system

ISO 9000 standards have 17 elements, which are management responsibility, quality system, contract review, product development, document control, purchasing, product identification and traceability, process control, testing and inspection, test equipment, inspection and test status, control of non-conforming products, corrective actions, handling, storage, packaging and delivery, quality records, internal quality audits, training, cleaning and disinfection and personal hygiene.

Management responsibility is the first and overall most important system requirement. Full commitment of the company top management is a must, and the entire system must be under management control and review. Management shall define the objectives and the policy of the system, and it bears the full responsibility for ensuring that the policy is understood, implemented and maintained at all levels in the company. Responsibility and authority of all personnel, who manage, perform and verify work affecting quality, shall be defined by the management, and adequate resources should be provided. If the HACCP concept is applied, it shall be stated in the company quality objective and policy. Quality system refers to the documented system ensuring that products comply with the specified requirements. It states that management shall ensure the presence of documented procedures and instructions in accordance with the ISO 9000 standard in question as well as efficient implementation of the Quality System Procedures and Instructions. The system is often organized in three levels comprising the Quality Manual, Procedures and Instructions. If the HACCP concept is incorporated with the more narrow quality objective, e.g. with *Salmonella* as the defined risk, only procedures and instructions for control of *Salmonella* as defined in the objective of the system will be included. It is laid down in the contract review that the producer shall review and evaluate all contracts to ensure that he/she can supply a product, which meets the customer's specified requirements and expectations, e.g. the product shall conform to specified requirements which for *Salmonella*, could be "absence in 25g frozen shrimps" in each of a certain number of packages according to the sampling plan agreed upon. Obviously, this element is very important in the ISO 9000 system designed to deal with supplier – customer relations. It is also stipulated that records of such contract reviews shall be maintained.

Under product development, the supplier shall establish and maintain procedures, which control and verify all phases of product development to ensure that the specified requirements are met. Using HACCP and *Salmonella* as an example, this means that the system shall ensure that new products and processes are not implemented unless they provide safety against *Salmonella* as laid down in the quality objective of the system. This is a complicated, demanding element of the standard, difficult to implement as well as maintain in the company. Documentation is a vital part of the system and so document control shall ensure that all necessary documents (procedures, instructions, forms, etc.) are available where needed, and obsolete documents promptly removed from all locations. It is a key point of ISO 9000 that purchasing (Requirement No.6) only takes place from approved suppliers, who have been selected on the basis of previous performance and an effective control system as well as their ability to meet the specified requirements. When applying the HACCP concept for the *Salmonella* in farmed frozen shrimp, it means that feed for the farm should only be purchased from feed mills producing feeds, which do not contain *Salmonella* according to the specifications agreed upon. The standard reaches even further in demanding mutual cooperation and a contractual understanding with the feed mill; the mill shall be assessed to be included in the list of approved suppliers established according to ISO 9000 requirements. The feed mill shall be audited just like all other suppliers on the list, and purchased products shall be inspected on-receipt, and feed-back on performance at all points shall be ensured. The very obvious reasons for these detailed requirements for purchasing obviously will be the inevitable effect of raw materials, machines, cleaning agents, services, etc. on the quality of the final product.

Producers for product identification and traceability during all stages shall be established, maintained and recorded. If required, each batch, package, etc. shall be provided with a unique identification, which shall be recorded. Process control shall ensure that all processes influencing the quality of the final product shall be specified and documented to ensure and verify that they are carried out under controlled conditions. This involves documented work instructions, including cleaning and disinfection procedures, use of appropriate equipment, machines, material and arrangement of processing facilities as well as monitoring of products and processes. This element is the key for the HACCP concept with the hazard analysis, identification of critical control points (CCPs) and monitoring of CCPs. A schedule for testing and inspection of raw materials, intermediate and final products shall be established. For the HACCP system, the schedule must be based on CCPs as identified in the hazard analysis. Test methods must be defined. Responsibilities for sampling and testing, reporting and control of non-conforming products shall be defined and reference made to the appropriate specifications. The test equipment used shall be selected to demonstrate acceptable compliance with the defined specifications for the products and shall be calibrated at regular intervals against nationally recognized standard references.

Proper identification of the inspection and test status shall be ensured with untested, tested, approved or rejected products being clearly marked. Procedures and instructions shall be established for control of non-conforming products. In the present example, shrimp containing *Salmonella* will be a non-conforming product according to the specifications agreed upon. Such a product shall be identified, placed and labeled in a way that clearly isolates it and prevents it from being supplied as *Salmonella*-free by mistake. The responsibility for making decisions on disposition of non-conforming products shall be defined and documented. A non-conformity report shall be worked out stating the nature of the non-conformity, the disposition decided, and the corrective action to the initiated for resolving the non-conformity as described in be following. The corrective action system is concerned with revising work operations. to try to eliminate the causes of failure. This is the system requirement that helps a company getting better and better by aiming at doing everything right the first time. To control all the activities required in the corrective action, forms containing the following points shall be applied: clear statements of the non-conformity, responsibilities, action to be taken, date of implementation, verification and recording of the resulting new procedures.

Handling, storage, packaging and delivery is obviously very important for foods, to prevent damage or deterioration of the products. Temperature control including monitoring and recording shall be mentioned as examples to illustrate the importance of this requirement, which obviously apply to all stages from raw materials, throughout production and delivery, and to the point of consumption. Determination and control of shelf life is needed, and so is full traceability with respect to the risk of product recall. Recording is required with the purpose of demonstrating the achievement of the required quality and to demonstrate that the quality system is effective. This is stated in quality records and the meaning of it is seen from the following examples of records to be included: inspection reports, analytical results, calibration reports, audit reports and corrective action reports.

It is also demanded that internal quality audits for the system be done on a regular basis. An appropriate audit plan shall be worked with ensuring that all elements are audited once a year. Audit teams shall be formed and it must be assured that the members are independent of the activities being audited. The audit report shall be included in the quality records as mentioned above. Management shall carry out its own independent review and evaluation of the quality systems. This has to be carried out on a regular basis twice per year and it should be based on the above international audit reports mentioned above as well as evaluation of the overall effectiveness of the system in achieving the quality objectives stated. Needs for updating, new strategies should also be indicated. It all has to be documented in a report. This again demonstrates the very active role to be played by the management of the company. Training is a vital part of the ISO 9000 standards. Of similar importance to food companies are cleaning and disinfection and personal hygiene. These subjects have been included as separate requirements to emphasize their importance and they have been used for the following illustration of the structure of the system with its various types of documents.

A company shall implement the following steps for obtaining registration as an ISO standard company.

1. Commitment by top management
2. Formation of ISO steering group
3. Training in ISO
4. Development of a quality manual
5. Start-up internal audits
6. Selection of a registrar
7. Pre-assessment
8. Improvement of system
9. Assessment
10. Surveillance

Documented quality system

ISO 9000 standards require a documented quality system. A three-level structure of documentation has proved effective in all the industries including food. Level 1 is described in the Quality Manual, which is normally a short easy-to read manual briefly stating the company's quality objectives and policies. All requirements of the appropriate ISO standard should be addressed. The Quality Manual need not contain confidential information, as it is intended for potential customers. Quality manual inspires the confidence of the customers. The individual pages of the quality manual shall be signed by the managing director or chairman of the governing board to show the commitment of the top management. Level 2 comprises procedures, describing how the statements of the quality manual are deployed and implemented in the company. Persons responsible, as well as where and when shall be stated. The procedures shall be issued by the Quality Control Manager and approved by the Technical Director. Level 3 contains the work instructions giving all details on how the contents of the procedures are accomplished. In Levels 2 and 3, appropriate references shall be given to various forms to be filled out including the list of approved suppliers mentioned earlier and being part of the system documentation.

Establishment and implementation of quality system

The establishment and implementation of quality system is a very demanding job both in terms of man-hours and resources. Proper planning including clearly defined project organization along with assistance from outside consultants is essential for a successful implementation. Full commitment and motivation as well as intensive training of all employees are also indispensable requirements. A Quality Management Group shall be formed for initiating the project as well as for its completion. In the case of food industries, this group comprises Managing Director, Technical Director, R & D Manager, Sales Manager and Head of the laboratory. Key functions of the group are definition of quality policy and objectives, definition of responsibilities, decision on time schedule for the project from start to certification, allocation of resources required, information and motivation of all staff members, training of all employees, follow-up on time schedules and resolving differences of opinions, arguments.

The time required for the implementation and certification of the system in a medium-sized company is 1 to 2 years. The various phases and activities after the formation of quality management group are hazard analysis, audit of present system elements, estimation of resources and total period of time required for the project including certification, formation of project organization, preparation of quality manual (Level 1), training of all staff concerned, definition of procedures and instructions (Levels 2 and 3) to be included in departmental manuals, i.e. table of contents, decision on time schedule for preparation of departmental manuals, establishment of working groups for the preparation of the individual procedures and instructions, commenting, reviewing, approving and issuing procedures and instructions, implementation of procedures and instructions, first approach to certifying body, internal auditing, corrections, adjustments, further training of staff and final certification.

The time schedule for establishing and implementing a quality system in a small-sized food processing plant is given below:

1. Hazard analysis	- 4	weeks
2. Audit of present system elements	- 20	weeks
3. Preparation of quality manual	- 6	weeks
4. Information and training of employees	- 5	weeks
5. Establishment of working groups, preparation of the procedures and instructions	- 30	weeks
6. Contact to certifying body	- 1	week
7. Adjustments, implementation, further training of employees, internal audits	- 10	weeks
8. Final adjustments	- 2	weeks
9. Certification	- 4	weeks
Total	82	weeks

Worldwide equivalents of ISO 9000 standards

Country	ISO 9000 standard nomenclature
Australia	AS 3900
Belgium	NBN-EN 29000
Brazil	NB 9000
Canada	Z 299
Chile	NCH- ISO 9000
China	GB/T 10300
Denmark	DS/ISO 9000
Finland	SFS-ISO 9000
France	NF EN 29000
Germany	DIN ISO 9000
Greece	ELOT EN 29000
Iceland	IST ISO 9000
India	IS 14000
Ireland	I.S/ISO 9000
Italy	UNI/EN 29000
Japan	JIS Z 9900
Mexico	NOM-CC 2
The Netherlands	NEN ISO 9000
Norway	NS ISO 9000
Portugal	EM 29000

[Table Contd.

Contd. Table]

Country	ISO 9000 standard nomenclature
Singapore	SS 306
South Africa	SABS 0157
Spain	UNE 66-9000
Sweden	SS-ISO 9000
Switzerland	SN-EN 29000
United Kingdom	BS 5750
United States	Q 9000

8.2.2. Codex Alimentarius Commission or Codex

The main international organization for food standards is the Codex Alimentarius Commission jointly brought out by the Food and Agricultural Organization (FAO) and World Health Organization (WHO). The aim of the Codex is to develop food standards to be used worldwide with a view to protect consumers' health and ensuring fair trade practices. Member countries use these standards as a basis to formulate their own standards. Codex documents include provisions in respect of good hygiene, food additives, contaminants, labeling and presentation and methods of analysis and sampling. The Codex publications are intended to guide and promote the elaboration and establishment of definitions and requirements in foods and to assist in harmonization of trade among countries by removing technical barriers. The Commission has established various specialist committees to deal with separate areas in food industry. They are of two types viz. Commodity Committees [e.g. Fish and fish products (Norway), Fats and oil (UK), Meat and meat products (Germany), Poultry and poultry meat (USA)] and General Subject Committees [e.g. Hygiene (USA), Labelling (Canada), Additives (Netherlands), Analysis and sampling (Hungary), General Principles on handling and processing (France)]. The Committee on fish and fish products holds its regular meetings in Norway. A number of standards on fishery products including the one on Quick Frozen Shrimps have been brought out by Codex.

8.2.3. United States Food and Drug Administration (USFDA)

The United States of America gives supreme importance to the safety of the general public and so enacted several laws. Laws enforced by USFDA in this respect are Food, Drug and Cosmetic Act, Public Health Service Act, Fair Packaging and Labeling Act Tea importation Act, Radiation Control for Health and Safety Act, and Import Milk Act. The laws and regulations are intended to ensure that the foods are pure, wholesome, safe and produced under sanitary conditions, that drugs and devices are safe and effective for their intended use, that cosmetics are safe and made from appropriate ingredients, and that all labeling and packaging are truthful, informative and not deceptive. The US Federal Food, Drug and Cosmetic Act prohibits distribution or importation of adulterated or misbranded articles in the US. The legal requirement of the imported and the domestic products are the same, but the enforcement procedures are different. Imported products regulated by the FDA are subject to inspection at the time of entry. Illegal or violative shipments are detained and destroyed or re-exported to the port of the origin. Regulations issued by USFDA are an important part of the Food and Drug Law such as Current Good Manufacturing Practice Regulations, New Drug Regulations and FDA Food Standards. The principal requirements of food law are health safeguards, economic safeguards, labeling requirements [including name and address of the manufacturer/ packer/distributor, net amount of food in the package, common name of the foods, ingredients in the foods, source of raw materials (in code form), nutrition labeling and production code indicating date of production, batches, etc.], sanitation requirements, personnel hygiene requirements, plant construction design and layout, defect action levels, food additives, warehousing, food standards and HACCP implementation.

USFDA Regulatory Requirements

Salmonella / Arizona	- Absent in 375 g
Listeria (cooked only)	- Absent
Staphylococcus aureus	- 100/g

Sulphur dioxide	- 100ppm
Mercury	- 0.5ppm
PCBs	- 2 ppm
DDT and its derivatives	- 5 ppm

Filth in fresh or frozen raw shrimp

A. Flies and other insects (whole or equivalent)

   1. Filth insect   - 2 in a sample

   2. Incident insect   - 3 in a sample

B. Insect fragments

   1. Filth insect fragments   - 5

      Fragments (excluding setae) in 2 of 6 subs

      (clearly identified as parts of the filth insect)

   2. Large body parts of filth insect

      (i.e. Thorax, abdomen)   - 1 in 2 of 6 subs

   3. Unidentified fragments   - 15 in a sample

C. Hairs

   1. Rat or mouse   - 2 of any size in a sample

   2. Striated but not of rat or - 3 of any size in a sample
      mouse

8.2.4. EU Hygienic Regulations

The Council of European Union prescribes stringent standards. The EU has specified 62 parameters (comprising 4 organoleptic parameters, 15 physico-chemical parameters, 24 undesirable substances in excessive amounts, 13 toxic substances and 6 microbiological parameters), besides 4 parameters for softened water as requirements relating to the quality of water intended for human consumption through its Council Directive No. 80/778/EEC. The important aspects of EU standards are listed below:

Microbial Standards for Aquatic food

Product	APC	FC	SA	SR	SAL	VC	LM
Chilled, quick frozen peeled & cooked	m =10,000 M = 1,00,000 n = 5 c = 2	m = 100 M = 100 n = 5 c = 2	m = 100 M = 1000 n = 5 c = 2	Absent in 25 g n = 5 c = 0	Absent in 25 g n = 5 c = 0	Absent in 25 g n = 5 c = 0	Absent in 25 g n = 5 c = 0
Chilled fish portions breaded or not	m =50,000 M = 5,00,000 n = 5 c = 2	m = 10 M = 100 n = 5 c = 2	m = 100 M = 1000 n = 5 c = 2	Absent in 25 g n = 5 c = 0	Absent in 25 g n = 5 c = 0	Absent in 25 g n = 5 c = 0	Absent in 25 g n = 5 c = 0
Frozen crustace-ans	m =50,000 M = 5,00,000 n = 5 c = 2	m = 100 M = 1000 n = 5 c = 2	m = 100 M = 1000 n = 5 c = 2	Absent in 25 g n = 5 c = 0	Absent in 25 g n = 5 c = 0	Absent in 25 g n = 5 c = 0	Absent in 25 g n = 5 c = 0

APC – Aerobic plate count/g; FC – Faecal coliforms/g; SA – *Staphylococcus aureus*/g; SR – Sulphite reducers/g; SAL – *Salmonella* in 25g; VC - *Vibrio cholerae*; LM – *Listeria monocytogenes*.

n – number of units comprising the sample; m – limit below which all results are considered satisfactory; M – acceptability limit beyond which the results are considered unsatisfactory; c – no of sampling units giving bacterial counts between M and m.

Heavy metal limits for aquatic food

SI. No.	Heavy metals	Limits (ppm)
1.	Cadmium	0.5 – 3.0 *
2.	Lead	0.5 – 10.0 *
3.	Zinc	50
4.	Copper	—
5.	Mercury	0.5 to 1.0
6.	Tin	250 ppm

*Varying between EU countries

Microbiological criteria (Cooked, ready-to-eat shrimp & crab meat)

Salmonella	- Not to be detected in 25 g
Staphylococcus aureus	- m = 100/g; M=1000
E. coli	- m = 10; M = 100
APC (shrimp without shell)	- m = 50,000; M = 5,00,000
APC (crabmeat)	- m = 1,00,000; M = 10,00,000

Quality requirements of water specified in EU Council Directive (80/778/EEC)

A. Organoleptic parameters

SI. No.	Parameters	Expression of the results	Maximum admissible concentration (MAC)
1.	Colour	mg/l Pt/Co scale	20
2.	Turbidity	Mg/l SiO_2 Jackson units	10
3.	Odour	Dilution number	2 at 12°C 3 at 25°C
4.	Taste	Dilution number	2 at 12°C 3 at 25°C

B. Physico-chemical parameters

SI. No.	Parameters	Expression of the results	Maximum admissible concentration (MAC)
5.	Temperature	°C	25
6.	Hydrogen ion Concentration	pH unit	Water should not be aggressive Maximum admissible value: 9.5

[Table Contd.

Contd. Table]

Sl. No.	Parameters	Expression of the results	Maximum admissible concentration (MAC)
7.	Conductivity	$\mu s\ cm^{-1}$ at 20°C	NS*
8.	Chlorides	Cl mg/l	NS*
9.	Sulphates	SO_4 mg/l	250
10.	Silica	SiO_2 mg/l	NS*
11.	Calcium	Ca mg/l	NS*
12.	Magnesium	Mg mg/l	50
13.	Sodium	Na mg/l	150
14.	Potassium	K mg/l	12
15.	Aluminium	Al mg/l	0.2
16.	Total Hardness		See Table F
17.	Dry residues	mg/l after drying at 180°C	1500
18.	Dissolved oxygen	$\%O_2$ saturation	Saturation value >75% except for underground waters
19.	Free carbondioxide	CO_2 mg/l	Water should not be aggressive

* NS – Not Specified

C. Parameters concerning substances undesirable in excessive amounts

Sl.No.	Parameters	Expression of the results	Maximum admissible concentration (MAC)
20.	Nitrates	NO_3 mg/l	50
21.	Nitrites	NO_2 mmg/l	0.1
22.	Ammonium	NH_4 mg/l	0.5
23.	Kjeldahl nitrogen (excluding N in NO_2 and NO_3)	N mg/l	1
24.	Oxidizability (K MnO_4)	O_2 mg/l	5
25.	Total Organic Carbon (TOC)	C mg/l	NS*
26.	Hydrogen sulphide	S µg/l	Undetectable organoleptically
27.	Substances extractable in chloroform	Mg/l dry residue	NS*
28.	Dissolved or emulsified hydrocarbons (after extraction by petroleum ether); Mineral oils	µg/l	10

[Table Contd.

Contd. Table]

Sl.No.	Parameters	Expression of the results	Maximum admissible concentration (MAC)
29.	Phenols (phenol index)	C_6H_5OH µmg/l	0.5 excluding natural phenols, which do not react to chlorine
30.	Boron	B µg/l	NS*
31.	Surfactants (reacting with methylene blue)	mg/l(lauryl sulphite)	200
32.	Other organochlorine compounds not covered by parameter No.55	µg/l	Haloform concentration must be as low as possible
33.	Ion	Fe µg/l	200
34.	Manganese	Mn µg/l	50
35.	Copper	Cu µg/l	NS*
36.	Zinc	Zn µg/l	NS*
37.	Phosphorus	P_2O_5 µg/l	5000
38.	Fluoride	F mg/l8-12°C25-30°C	150070
39.	Cobalt	Co g/l	NS*
40.	Suspended solids		NS*
41.	Residual chlorine	Cl µg/l	**
42.	Barium	Ba µg/l	NS*
43.	Silver	Ag µg/l	10

* NS – Not Specified

** Necessary measures should be taken to ensure that substances used in the preparation of water for human consumption do not constitute a public health hazard

D. Parameters concerning toxic substances

Sl. No.	Parameters	Expression of the results	Maximum admissible concentration (MAC)
44.	Arenic	As µg/l	50
45.	Beryllium	Be µg/l	NS*
46.	Cadmium	Cd µg/l	5
47.	Cyanides	CN µg/l	50
48.	Chromium	Cr µg/l	50

[Table Contd.

Contd. Table]

Sl. No.	Parameters	Expression of the results	Maximum admissible concentration (MAC)
49.	Mercury	Hg µg/l	1
50.	Nickel	Ni µg/l	50
51.	Lead	Pb µg/l	50(in running water)
52.	Antimony	Sb g/l	10
53.	Selenium	Se µg/l	10
54.	Vanadium	V µg/l	NS*
55.	Pesticides and related products 1. Substances considered separately 2. Total	µg/l	0.10.5
56.	Polycyclic aromatic hydrocarbons	µg/l	0.2

*NS – Not specified

E. Microbiological parameters

Sl. No.	Parameters	Expression of the results	Maximum admissible concentration (MAC)	
			Membrane filter method	Multiple tube method (MPN)
57.	Total coliforms	100	0	MPN<1
58.	Fecal coliforms	100	0	MPN<1
59.	Fecal streptococci	100	0	MPN<1
60.	Sulphite-reducing Clostridia	20	–	MPN1

Sl.No.	Parameters	Results: Size of sample in ml		Maximum admissible concentration (MAC)
61.	Total bacterial counts for water supplied for human consumption	37°C	1	NS*
		22°C	1	NS*
62.	Total bacterial counts for water in closed containers**	37°C	1	20
		22°C	1	100

* NS – Not Specified

** Values should be measured within 12 h of being put into closed containers

F. Minimum required concentration for softened water intended for human consumption

Sl. No.	Parameters	Expression of the results	Maximum admissible concentration (MAC)
1.	Total hardness	mg/l ca	60
2.	Hydrogen ion	pH	Water should not be aggressive
3.	Alkalinity	mg/l HCO_3	30, Water should not be aggressive
4.	Dissolved oxygen	–	Water should not be aggressive

A compilation of seafood quality criteria by countries such as USA, EU and Japan are given below:

Standards for import of fish / fishery products

Raw Products		
1.	TPC at 37°C	:	5×10^5/g
2.	Faecal coliforms	:	20/g (by MPN method)
3.	E.coli	:	20/g (by MPN method)
4.	Coagulase +ve Staphylococci	:	100/g
5.	Salmonella	:	Absent in 25g
6.	Vibrio cholerae	:	Absent in 25g
7.	Listeria monocytogenes	:	Absent in 25g
8.	Histamine	:	≤ 50 ppm
9.	Indole	:	< 50 microgram/100g
10.	TVB-N	:	< 30 mg/100g
11.	PSP	:	80 microgram/100g
12.	DSP	:	20 microgram/100g
13.	Filth	:	Nil
14.	DDT/DDE	:	< 5 ppm
15.	Aldrin, Dieldrin	:	< 0.3 ppm
16.	PCB's	:	< 2 ppm
17.	2,3,7,8 TCDD (Dioxin)	:	20 ppt
18.	Cd	:	< 3 ppm
19.	Pb	:	< 1 ppm
20.	Hg	:	0.5 ppm
21.	Antibiotic residues	:	Nil (0.1 ppm tetracylines by Japan)

Cooked products		
1.	TPC at 37°C	:	1×10^5/g
2.	*E.coli*	:	< 1/g (by MPN method)
3.	Faecal coliform	:	< 1/g (by MPN method)
4.	Coagulase +ve Staphylococci	:	100/g
5.	*Salmonella*	:	Absent in 25g
6.	*Vibrio cholerae*	:	Absent in 25g
6.	*Listeria monocytogenes*	:	Absent in 25g
7.	Histamine	:	$\leq$ 50 ppm
8.	Indole	:	$\leq$ 50 microgram/100 g
		:	$\leq$ 25 microgram/100 g (shrimp)
9.	PSP	:	80 microgram/100 g
10.	DSP	:	20 microgram/100 g
11.	Filth	:	Nil
12.	DDT / DDE	:	< 5 ppm
13.	Aldrin & Dieldrin	:	< 0.3 ppm
14.	PCB's	:	< 2 ppm
15.	Cd	:	< 3 ppm
16.	Pb	:	< 1 ppm
17.	Hg	:	0.5 ppm
18.	Antibiotic residues	:	Nil (0.1 ppm tetracylines by Japan)

REFERENCES

1. Ababouch, L., Gandini, G. and Ryder, J., 2005. Causes of detentions and rejections in International fish trade.

2. Bonnell, A.D., 1994. Quality assurance in seafood processing: A practical guide

3. CAC, 2001. Codex Alimentarius Commission Food hygiene basic texts. 2nd Edition

4. Cesarettin, A., Fereidoon, S., Kazuo, M. and Udhaya, W., 2011. Handbook of Seafood Quality and Safety and Health Applications

5. Connell, J.J., 1995. Control of fish quality

6. EC, 1991. European Council Directive 91/492/EEC of 22 July 1991 laying down the health conditions for the production and the placing on the market of fishery products

7. Export Inspection Council of India. https://eicindia.gov.in/

8. FAO/WHO, 1978. Recommended International Code of Practice for frozen fish

9. FAO, 1998. Food quality and safety system - A training manual on food hygiene and the hazard analysis critical control point (HACCP) system

10. Food Safety Standards Authority of India. https://fssai.gov.in/home

11. Haynes, P.R., 1995. Food microbiology and hygiene

12. Huss, H.H., 1988. Fresh fish – Quality and quality changes

13. Huss, H.H., 1994. Assurance of seafood quality

14. Huss, H.H., Jacobsen, M. and Liston, J., 1992. Quality assurance in the fish industry

10. Huss, H.H., 1995. Quality and quality changes in fresh fish

11. Huss, H.H., Ababouch, L. and Gram, L., 2004. Assessment and management of seafood safety and quality

15. Huss, H.H., 2003. Assessment and Management of Seafood Safety and Quality. FAO Tech. Paper No. 444.

16. ICMSF, 1988. Microorganisms in foods. 4. Application of the hazard analysis critical control point (HACCP) system to ensure microbiological safety and quality

13. ICMSF, 1996. Microorganisms in foods 5. Characteristics of microbial pathogens

14. Iyer, T.S.G., Kandoran, M.K., Thomas, M. and Mathew, P.T., 2002, Quality assurance in seafood processing

15. Jouve, J.L., Stringer, M.F. and Baird-Parker, A.C., 1998. Food safety management tools

16. Kramer, D.E. and Liston, J., 1986. Seafood quality determination

17. Leo, M.L. and Fidel, T., 2010. Handbook of Seafood and Seafood Products Analysis

18. Lund, B.M., Parker, B.T.C. and Gould, G.W., 2000. The microbiological safety and quality of foods

19. Marine Products Export Development Authority of India. http://mpeda.gov.in/MPEDA/

20. Naaum, A. and Hanner, R., 2016. Seafood Authenticity and Traceability

21. Pearson, A.N. and Dutson, T.R., 1995. HCCP in meat, poultry and fish processing

22. Price, R.J. and Tom, P.D., 1999. Compendium of fish and fishery product processes, Hazards and Controls

23. USFDA, 1995. Procedures for the safe and sanitary processing and importing of fish and fishery products: Final Rule. Code of Federal Regulations, Parts 123 and 1240. Volume 60. No.242

24. USFDA, 2000. United States Food Safety System. Published online at http://www.foodsafety.gov./-fsg/fssyst2.html

25. USFDA, 2001. Fish and fishery products hazards and controls guide. 3rd Edition

26. Vincent, K. and Omachonu, J.E.R., 2004. Principles of Total Quality

27. Ward, D.R. and Hackney, C., 1991. Microbiology of marine food products

28. WHO, 1995. Food safety issues: Food technologies and public health

29. WHO, 1999. Joint FAO/NACA/WHO Study Group on food safety issues associated with products from aquaculture - WHO Technical Report Series No.883